Cockroaches or R

Cockroaches care, facts and information.

Including German Cockroach, American Cockroach, Madagascar Hissing Cockroach. Keeping, breeding, health all covered.

by

Elliott Lang

Published by IMB Publishing 2013

Table of Contents

Table Of Contents

Table Of Contents

Foreword

When it comes to looking for a pet that is intriguing and easy to care for, it is well worth your while to consider cockroaches!

Many people automatically have a flinch reaction when it comes to these fascinating insects, but the truth is that this reaction is largely undeserved. While there are a few cockroaches that are harmful to humans and spread disease, the truth of the matter is that most cockroaches are simply animals that live in woods, fields, or deserts and do not bother humans at all.

Over the last twenty years or so, more and more people are realizing that these animals make excellent pets. They are clean, they come in a wide variety of shapes and sizes, and they are remarkably easy to care for.

Some people also raise cockroaches as a food source for their other pets, realizing that they are cheaper to raise and easier to breed than crickets. Cockroaches are even eaten by people!

Before you pass up on a great pet, learn more about cockroaches and what place they might have in your lif

Chapter 1) General Information

1) What is a cockroach?

The term "cockroach" covers a number of different insect species. On the whole, however, a cockroach is an insect that can range from less than half an inch long to more than three inches long. It typically has a long, narrow oval in shape, and it has a relatively sleek body. Like all insects, it has six legs, a segmented body and antennae, which it uses to explore the world around it.

Cockroaches have eyes that are fairly good at distinguishing dark from light, but their real skills have to do with their sense of smell. Their sense of smell is far more powerful than a human's, and it allows them to track food across comparatively long distances.

Cockroaches may have wings that are long and translucent, or they may have wings that are short and stubby. Some cockroach species lack wings at all!

Cockroaches are cold-blooded, meaning that they need a heat source around at all times. In the wild, they rely on natural heat sources to keep them warm and to help them digest their food. In captivity, it is important to keep cockroaches at a warm temperature in order for them to thrive and breed.

Cockroaches are, on the whole, very gregarious animals. They like to congregate in one spot with others of their species, and there is even some evidence that cockroach mothers tend their young.

Some cockroach species are serious pests in human habitations. They get into food sources, and if left unchecked, they can carry diseases to people and to pets. They are often associated with disease, filth and poor housing conditions, but though it is true that these buildings are more prone to cockroach infestations, it is also true that any house can suffer these nuisances.

There are more than 4000 species of cockroaches around the world, and there is fossil evidence of cockroach-like species that go back more than 320 million years, making them one of the oldest living species that is still around and functional in the modern day. Even more interestingly, they have not changed much at all in that time period. A cockroach found in amber was distinctly recognizable as a cockroach even though it had been hatched millions of years before.

Despite their troublesome associations with poor housing and poor care, they are fascinating insects, and many people are realizing that they make excellent pets as well.

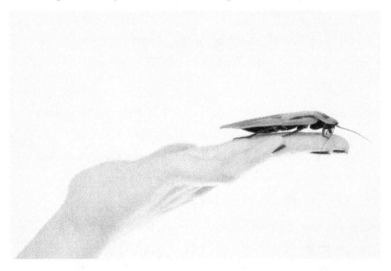

2) Life Cycle

In the wild, cockroaches are typically prey animals. That means that they tend to live very short lives, even when things are going relatively well.

Female cockroaches, depending on the species, can produce as many as 30 young in a single egg sac, and if the conditions are good, they can create several egg sacs in one year.

Most cockroaches tend to live between one to three years in the wild, but cockroaches that live in captivity can live much longer.

On top of that, the size of the cockroach also seems to determine its lifespan. Larger cockroaches can live to be between three to five years with one or two species living as long as ten years.

Ideal conditions such as good temperature, good humidity, and plentiful food can help a cockroach live longer. With good care, your pet cockroach can be a companion for many years to come.

It is also worth noting that cockroaches that live alone, even if it is a little unnatural for them to do so, often live longer than cockroaches that live communally. It is not natural for cockroaches to live alone, but when they do so, they seem to be free of some of the damage that goes with living with others.

3) In the Wild

Given the fact that there are more than 4000 species of cockroaches in the wild, it is difficult to give an accurate viewpoint of what an "average" cockroach's life might be.

Whether they live in the desert or the rain forest, however, cockroaches are fond of dark places with some humidity. They often live together in loose family clusters, and they are nocturnal, though there are many exceptions.

When it comes to food, some species are almost entirely herbivorous, but most species are simply opportunistic feeders, meaning that they will eat anything that crosses their path. They are not hunters; instead, they are largely scavengers, eating food that they find or food that larger animals have left on their way. There are quite a few species that love to eat rotten wood. In many ways, they serve well as decomposers in their ecological niche.

Cockroaches that live in human habitation as parasites and pests are also considered to be wild cockroaches. Though they live in the house, they are not cared for or husbanded by people. Instead, they treat the entire house as their own natural environment, creating a situation where they are much at odds with the homeowners!

Originally, cockroaches that were known to live in human habitations lived in the wild, but over time, it turned out that human habitations were adapted to their needs. Cockroaches that live in houses are drawn to warmth, and are often more tolerant of dry conditions than their cousins living in the fields, and also quite small and flat, meaning that they could easily fit through the crevices and cracks in the home.

They are found all over the world in every environment, except for the polar landscapes, and they are extremely enduring. The trait that makes them a difficult pest to get rid of in human habitation is actually one that makes them quite durable in the wild.

For the most part, cockroaches in the wild die long before they live out their natural span. Cockroaches in captivity live much, much longer, while cockroaches in the wild need to deal with predators, disease and simply poor luck.

One interesting fact is that cockroaches will die over time if they are flipped on their backs. Cockroaches cannot right themselves if they are turned on their backs, and once flipped, they will die after enough time. There is even a pesticide that is designed to cause spasms that result in being upended in just that way.

4) Predators

In the wild, there are many things that eat cockroaches. They are quite vulnerable, and given that their survival strategy is to simply have many young in the hopes that they survive, there are many cockroaches that are devoured along the way.

They are very prone to predation by birds, and there are even some species who hunt cockroaches specifically. These birds stalk cockroaches that hide in rotting logs and either pry them out with long beaks or simply peck at them as they trundle around.

Because cockroaches are very associated with water, they are also good prey for fish and amphibians of all types.

For the most part, cockroaches are only preyed upon by small predators. Most larger predators find cockroaches to be too little food for the trouble.

However, animals like cats and very young predators might catch a cockroach in order to play with it or to learn to hunt. If you have cockroaches in your home, you may notice that your cat or dog eats one once in a while.

5) Maturation

The life of a cockroach can range anywhere from one year to as much as ten years, but in general, all cockroaches have the same life cycle.

After mating, the female cockroach carries its eggs in an egg sac known as the ootheca. The ootheca is made from a type of protein that is designed to protect the tiny eggs inside. Some cockroaches, like the popular pet species the Madagascar hissing cockroach, carry the ootheca on the underside of the abdomens. Other species carry the ootheca inside their bodies, resulting in the birth of live young.

There is only one species that carries the eggs inside the body without the ootheca. This species is considered to be viviparous; giving birth to live young in much the same way that a mammal would.

In some cases, the mother cockroach hides the ootheca or buries it in the ground. After that, she leaves and the baby cockroaches, when they hatch, are completely independent. They are on their own in the face of predators, operating entirely on instinct. In these species, there is no recognition between parent and child.

On the other hand, however, there are species where the mother carries the ootheca either internally or externally until the eggs hatch. Then she cares for the young cockroaches, known as nymphs, and protects them for a short while. She shelters them from predators and even teaches them how to find food. In some cases, she will bring them their first meals. In these species, scientists theorize that there is at least some recognition between mother and child.

When they hatch, young cockroaches are usually white. Within the span of a few hours, however, they darken to a black or brown, which is usually much more useful for hiding from predators. Most cockroach species mature within a few months, and as they grow, they shed their skin, leaving the molts behind. Shedding, scientifically known as ecdysis, is a process where an invertebrate grows too large for its skin. In cockroaches, this usually means that the old skin becomes white and brittle, allowing the cockroach to split out the back of the shell to emerge with a fresh new skin underneath. For a few hours after shedding, the cockroach's shell is soft, but in a short amount of time, it will harden.

Depending on their species and their size, cockroaches might shed as many as five or six times during their lives. They will grow more, and thus molt more, when there is plenty of food around. If your cockroaches are shedding regularly and easily, you will find that they are usually fairly healthy.

On the whole, cockroaches are gregarious. Though they can be kept singly and in pairs as pets, in the wild they tend to live in colonies. However, while they are gregarious, they are not as social. Unlike bees and termites, where different individuals fulfill different roles, cockroaches have no structure.

Instead, each insect adds its intellect and information to the whole, which leads to a certain type of group-think on behalf of the colony. This is evidenced in the fact that so long as a certain space is large enough for the colony, they will stay. However, if that space shrinks or there are more cockroaches than it will hold, small groups will sheer off from the whole.

Some scientists have actually managed to affect the behavior of cockroach groups by placing a robotic cockroach in the middle of a group of cockroaches. Through giving off cues and making decisions, the fake cockroach could cause the group to do things that they would not otherwise do, like run towards lighted areas.

As soon as cockroaches are fully-grown, they are capable of breeding, and in many cases, it takes a very small amount of time. Some cockroaches mature within a month or two, with most taking three or four months before they are ready to produce young.

The males fight amongst themselves for the right to mate with the females, and these fights are usually fairly sedate. One male will push another in a contest of strength, and in many cases, there is no injury at all. Once in a while, there will be signs of more serious aggression, but increased aggression in cockroaches is frequently associated with other changes as well.

In some cockroach species, it is fairly easy to tell the males from the females, while in others, it is next to impossible.

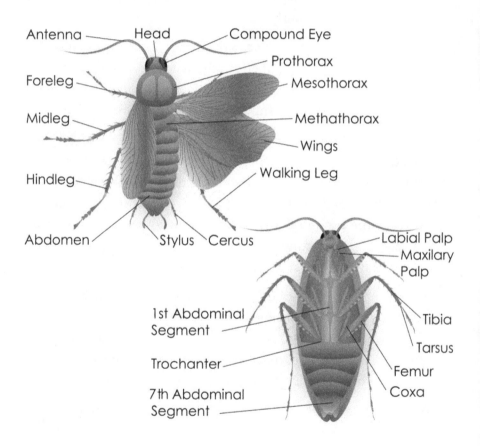

Chapter 2) Cockroach Myths

Cockroaches are animals that have been a part of human history since the beginning. Given this fact, it makes a great deal of sense that there would be plenty of myths about them, and also plenty wrongful information. Check your cockroach knowledge to see where you stack up.

MYTH: Cockroaches would survive a nuclear explosion.

TRUTH: Cockroaches will survive a nuclear explosion better than humans, but they will still be harmed.

Cockroaches are similar to other insects in that they are less susceptible to radiation than humans are. In fact, their resistance may be 10 times as good as that of humans to radiation. However, that does not mean that they are immune to harmful effects from a nuclear explosion.

MYTH: Cockroaches are only drawn to dirty homes.

TRUTH: Cockroaches can be found everywhere.

Too many people associate cockroaches with poverty and poor housekeeping. The truth of the matter is that cockroaches are quite adventurous, and they can be found just about everywhere. They can hitchhike in luggage or on clothing, and they can come into the house from the wild.

MYTH: You can get rid of cockroaches by bringing in their natural predators.

TRUTH: It is far better to bring in exterminators.

There are plenty of things that eat cockroaches in the wild, but the cockroach is designed to be very resistant to getting wiped out entirely. Just because there are a wide number of animals that eat cockroaches, and even parasites that prey on cockroaches' egg

17

sacs, you will find that these predations do little to reduce the number of cockroaches in the population. Do not allow your pet to eat wild cockroaches in an attempt to control them!

MYTH: All cockroaches can fly/climb/jump.

TRUTH: Different cockroaches have different features.

It is hard to look at one cockroach and say that this is what all cockroaches are like. There are the large and docile hissers, there are the insidious German cockroaches and there are plenty of exotic cockroaches that look different from both. As a matter of fact, there are cockroaches that fly, jump, climb, do all three or do none of these things. Each species is adapted well for its environment and its ecological niche.

MYTH: Cockroaches fear light.

TRUTH: Many cockroaches are nocturnal.

There are several iconic movie scenes where a light is turned on and the cockroaches dash for cover. This is something that has given many people the idea that cockroaches fear light.

The truth is that when cockroaches scuttle away from the light, they are simply startled. They respond to any changes in the environment that might signal a predator. Chances are good that if you came across a cockroach trundling around in the day light that you could startle it just as easily by suddenly casting a shadow over it!

Most of the cockroaches that end up in human habitations are very nocturnal, which often contributes to this myth.

MYTH: Cockroaches are dirty.

TRUTH: Most cockroaches keep themselves quite clean.

If you have cockroaches, one of the first behaviors that you will

18

notice is grooming. These insects can seem quite fastidious when it comes to preening their feet and their antennae. Grooming keeps the cockroach sharp and allows it to stay in good shape. Cleaning its sensory organs makes it more alert.

The issue is that while cockroaches themselves are clean, they often crawl through things that are not. On top of that, given the fact that the vast, vast majority of cockroaches do not end up in human habitation, they mostly avoid things like sewage and human filth.

Chapter 3) Types of Cockroach

With more than 4000 cockroach species to choose from, a complete list of these fascinating insects would take some time! Instead, check this list for cockroaches that are pests in human habitations, cockroaches that are almost only found in the wild, cockroaches that are kept as pets, and cockroaches that are often kept as feeders.

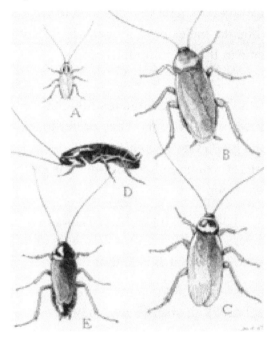

Image:

A. German Cockroach

B. American Cockroach

C. Australian Cockroach

D. Oriental Cockroach, female (wingless)

E. Oriental cockroach, male (winged)

21

1) American Cockroaches

Type: Pest

American cockroaches are among the largest cockroaches that regularly infest human dwellings. Both males and females can grow to be between one and two inches long, and they are both rather slender with dark brown bodies. The females tend to have a rather swollen abdomen, while the males have longer wings. Both males and females, however, can fly and jump fairly long distances.

These cockroaches are known for infesting large institutional buildings, apartments and sewer-related areas. They prefer to stay in the dark and the damp, and given that a female can produce somewhere in the vicinity of 150 young in her lifetime, it is easy to see how an infestation can occur.

From egg to adult, an American cockroach can live for just around 3 years, but this is a very long lifespan for members of this species. In most cases, they tend to die of accident or disease before they get that old.

These cockroaches are associated with disease, and they have a strong association with places that store human waste. Because of this, they are considered a nuisance species that can be harmful to people that share space with them. They spread bacteria and toxins when they travel through human habitations.

They are opportunistic feeders, and they seem to gravitate towards sweet tastes. They will happily eat fruits and vegetables, and they have been shown to have a marked preference for sugar, water and candy. However, when they are hungry, they will eat just about any kind of organic matter, including things like hair and decaying trash.

When you want a pet cockroach, do not simply capture an American cockroach. It is far better to purchase a cockroach from a responsible breeder who keeps their specimens in sterile conditions.

2) German Cockroaches

Type: Pest

German cockroaches are the pests that most people think about when they are dealing with household pests. These cockroaches are small, measuring only a little over half an inch, but they are pervasive and numerous. Unlike American cockroaches, which are only infrequently found in actual houses, German cockroaches are much more common.

German cockroaches are the most common roach to infest human habitations, and on top of that, they are among the most difficult to get rid of. They are incredibly hardy, and they have proven immune to many of the treatments that are used to get rid of similar pests.

They nest in human homes by preference, and though they only live to be between two and three years old, they can produce a high number of young. This, combined with their unending search for warmth and food, can make them a serious health hazard in apartments, in houses and even in institutional settings like libraries, shelters and government buildings.

Like American cockroaches, they are opportunistic feeders. Most exterminators caution that when you can see one, there are dozens that you can see. If you spot a German cockroach in your home, take the time to catch it and call an exterminator for help!

There is also a species that is known as the Asian or oriental cockroach, and though they differ in a few small ways, like the shape of their wings and the narrowness of their bodies, they are functionally identical to German cockroaches. The extermination methods are the same, and the infestation issues are the same as well.

3) Wood Cockroaches

Type: Wild

The wood cockroach, more commonly called the wood roach, is a forest-dwelling insect, and it is one of the vast number of cockroaches that do not really trouble human beings. Instead, they like to spend their time in very moist environments, and they are frequently found in wooded areas near lakes, rivers and streams.

In appearance, they are quite similar to the American cockroach. They are light to dark brown in color, and they are typically just over one inch long.

On occasion, wood cockroaches will wander into a human habitation, but fortunately enough, this is strictly a temporary problem. They will quickly wander right out again as soon as they find their way, and more importantly, they will not breed in your home. The wood cockroach requires a great deal of moisture before it can mate or lay eggs, and that means that it needs to stick close to outdoor sources of water.

If you are wondering whether a cockroach is an American cockroach or a wood cockroach, the thing to watch for is the activity level. American cockroaches are nocturnal and shun light. If you turn on the light, they will generally scatter and make a break for the darker cracks of the room.

A wood cockroach, on the other hand, is much more bold in the light. They are diurnal, and they will happily root around looking for food even when the light is on.

While wood cockroaches are not a huge house pest, it is worth looking at how they can get into the house. For the most part, these insects are brought in as hitchhikers. If you go camping, check your gear for any stowaways before you put it away in your home.

Also, if you bring firewood from the outside, inspect the logs carefully before you store them in the house. When you store your wood outside, it is best to keep it at least 15 feet away from the eaves of your house if you can.

Some people stack their wood close to the house to keep it dry and keep it accessible, but this gives insects like cockroaches, termites, ants and ticks a chance to gain entry to the inside of your home through the cracks. Instead, stack the wood a good distance away, and simply keep it covered.

4) Madagascar Hissing Cockroaches

Type: Pet

When it comes to a pet cockroach, the first one that comes to mind should definitely be the Madagascar hissing cockroach. This cockroach is large and slow-moving. In general, it is quite docile. It does not bite at all, and it is an ideal first pet for any adult or child who wants to try their hand at raising bugs.

The Madagascar hissing cockroach ranges between 2 and 3 inches in length, and as one of the larger cockroaches around, it weighs just a little shy of an ounce. There is a fair amount of heft to this insect, meaning that it is easier to handle.

On top of that, the Madagascar hissing cockroach is wingless. Unlike other cockroaches, it does not fly. However, it is worth noting that it does have sharp spikes on its legs, which are largely used defensively. If you grab at it carelessly, you may get a cut or two.

In the wild, these cockroaches are largely herbivorous, but like just about all other cockroaches, they have protein needs that are easily met with dry kibble.

In the pet trade, Madagascar hissing cockroaches are often called hissers, a name they share with several other species. They produce this noise by forcing air through their spiracles, which are organs that are located on each segment of their abdomen. When they do this, they make an aspirated hiss that is quite loud if you are standing close.

Hissing serves several important purposes in the lives of these cockroaches. Firstly, the hiss can be used to warn other cockroaches that there is danger approaching. These cockroaches live in large colonies, and they do look out for one another.

Hissing is also used defensively. It is a sound that can startle off predators and make them think twice about trying to turn this cockroach into a meal.

Finally, the hissing sound is also used to show dominance. When males are fighting for territory, for food or for mating privileges, they force air through their spiracles as loudly as they can.

Fighting males shove each other with the short horn-like structures on their head, making smaller and weaker males give way. There is a correlation between hissing and dominance; dominant males seem to hiss more frequently than submissive males across the board. Madagascar hissing cockroaches live for between two and five years. This makes them a great pet in many ways because they do not die as quickly as some other types of insect pets out there. They hang around for a while, and if you have just a few, you can observe their personalities and their inclinations.

Like the vast majority of cockroach species, the Madagascar hissing cockroach is not considered a pest for humans at all. In the wild, they are not drawn to human lodgings, and instead mostly spend their time searching out decaying plant matter and fruit. They typically live in rotting logs. Madagascar hissing cockroaches are a fantastic choice when you want to get started with keeping cockroaches as pets. They can be bred as feeder animals for your lizards, snakes or fish as well, but there are other species that produce young more readily as well.

Males will fight for dominance and the right to mate with the females, and once one male comes out on top, he will back his abdomen up to the female's abdomen, coupling with her while both their heads are facing different directions. This seems to be unique among the world of cockroaches. Madagascar hissing cockroaches give birth to live young in that they carry the ootheca inside their bodies until the young are ready to emerge.

5) Tiger-Striped Hissing Cockroaches

Type: Pet

The tiger-striped hissing cockroach is a popular pet variety of cockroach that is very similar in care and temperament to the Madagascar hissing cockroach. Unlike the Madagascar hissing cockroach, however, which has a body that is typically a rich brown, the tiger-striped hissing cockroach has an abdomen that is marked with alternating stripes of black and pale yellow or white. This makes this specimen quite distinctive and attractive, and it does very well as an ornamental pet.

The tiger-striped hissing cockroach is one of the largest roaches around, and it ends up being just a little longer than the Madagascar hissing cockroach, though they are similar in weight.

It is worth noting, however, that where the Madagascar hissing cockroach is rather slow and easy to catch, these cockroaches are a bit faster. Tiger-striped cockroaches can move quickly when they hit the ground, so people who are interested in keeping them should take extra care that they do not escape.

6) Harlequin Cockroaches

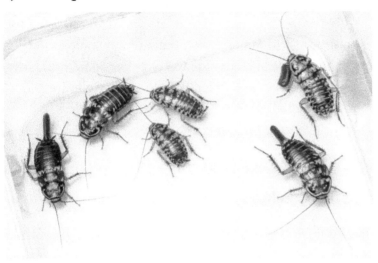

Type: Pet

When you are looking for a truly impressive cockroach specimen to keep as a pet, consider keeping harlequin cockroaches. These cockroaches are small, only going over an inch in length on occasion, but their variegated pattern makes them stand out from the duller Madagascar hissing cockroaches and even the tiger-striped hissing cockroach. Their checkered black and yellow bodies give them a look that is similar to a jester's uniform.

These cockroaches have the interesting trait in that they produce a sweet chemical that smells a little like pear drops. This chemical is slightly toxic, and it is intended to help them warn off predators. This means that the harlequin cockroach is a poor choice when you are looking for a cockroach that you can feed to your pets.

It is also worth noting that harlequin cockroaches are very fast.

Once they hit the floor, they can practically disappear. This species is not known for calm or steady movement, so take some care if you choose to keep them.

This is a lovely pet to have, but between its speed and its need for high temperatures and high humidity compared to some easier species, it might be best off for someone who has some previous experiences with cockroaches. Once you do get harlequin cockroaches established, however, they are good breeders, with the females producing up to 30 eggs in one egg sac.

7) Zebra Cockroaches

Type: Pet

With their distinctive black and white-striped abdomens, it is easy to see where zebra cockroaches get their name. This smaller cockroach only grows to be a little more than an inch in length, but it is an excellent choice for someone who is looking for a cockroach that is visually distinctive.

This is not a cockroach that flies, but it is an excellent climber. This means that if you choose to keep this cockroach, your container must have a lid, and ideally, it will also have a few inches of Vaseline smeared around the edges to prevent the cockroaches from climbing to the lid.

8) South American Firefly Beetle Mimic Cockroaches

Type: Feeder and Pet

When you are looking for something very special, make sure that you consider trying to locate a South American firefly beetle mimic cockroach, more often called a firefly roach or a firefly mimic.

These cockroaches are fairly small, ranging from between half an inch to one inch. The females are usually twice the size of the males.

While these cockroaches do not glow, they resemble fireflies to an impressive degree; their lower abdomens are even a paler color to mimic the firefly's luminescence. This small cockroach does fly and climb, so it is worth taking a great deal of care when it comes to keeping them contained.

They can be used as feeders, though they are a little more obscure than the dubia cockroach and other common feeder species. The key to breeding this southern cockroach is to get the heat and humidity up to fairly high and then to make sure that you provide them with a loose substrate to burrow in.

Females of this species do lay eggs, but instead of carrying the eggs or depositing them in another location, they carry the ootheca inside their bodies. As a result, it looks like they are giving live birth when they expel their young.

9) Giant Cave Cockroaches

Type: Pet

If you want something relatively large and fairly impressive, you cannot go wrong with the giant cave cockroach. This type of roach is one of the largest in the world, and it is often more than three inches long and an inch and a half wide. Despite its size, it is very thin, unlike hissers, which tend to look rather plump.

This is a fairly simple cockroach to find on the pet market. Originally, giant cave cockroaches were a great choice for scientists who were studying insects, and now they are being bred by hobbyists. They are simple to breed, though sexing these creatures takes a little more patience than it does for other cockroaches.

To sex giant cave cockroaches, flip them on their backs and look for two antennae-like structures on the last segment of their abdomen. These structures are known as cerci, and between them, males have a thread-like appendage called a stylus. Females lack this appendage.

This is a cockroach that does well in temperatures between 70 and 80 degrees, and they will breed prolifically if the enclosure is kept fairly humid and in the upper 70s.

10) Orange Head Cockroaches

Type: Feeder and Pet

The orange head cockroach hails from South America, and this brightly colored insect is kept both as a feeder and as a pet. It is a good-sized cockroach that tops out at around 2 inches, but the most colorful specimens tend to be less than half that. The nymphs are notable for being a rather velvety deep brownish-red instead of white or yellow. Until they are full adults, the nymphs spend most of their time burrowing in any available substrate.

It is worth noting that if this is a species that interests you, you should remember that they are somewhat aggressive. Unlike other cockroaches that are relatively laid back and quite docile, orange head cockroaches are known inside the pet trade for biting off each other's wings. It is very common to see an orange head cockroach with ragged wings or that is missing wings entirely.

They cannot be kept with any other cockroaches at all, as they will eat them. This is very important if you want to maintain more than one colony of cockroaches. The orange head cockroach is highly aggressive towards other species.

There is some evidence that states that if enough protein is available, the incidents of aggression go down. It might be worth boosting this cockroach's diet to 60 percent protein over the 40 or 50 that is typically recommended.

One thing that makes this cockroach so popular is that it cannot climb glass at all, making it a good choice for open terrariums.

Picture: Red Cockroach

11) Rhinoceros Cockroaches

Type: Pet

When you are looking for an excellent pet cockroach that is going to be quite impressive, look around for breeders selling rhinoceros cockroaches. These are some of the largest cockroaches in the pet trade, and they can grow to be more than 3 inches long.

This species does not have wings, and it is quite heavy for an insect. This species originally comes from Australia, where it usually contents itself with dead eucalyptus leaves. In captivity, they can eat the same kibble and fruit as other cockroaches.

One of the most interesting features of the rhinoceros cockroach is that it has tiny shovel-shaped structures on its front feet, allowing them to dig in the ground for the food that they prefer.

Another reason to get the rhinoceros cockroach is that it is an extraordinarily long-lived cockroach. Unlike other species that might get to three years if they are lucky, this cockroach can live to be upwards of ten years old. This insect is not even fully developed until it is around five years old.

Due to the long maturation period, do not get this cockroach if you are interested in being a breeder. People that breed this species need to be very patient and very willing to work with its particular nature and longevity.

12) Death's Head Cockroaches and Hybrid Death's Head Cockroaches

Type: Feeder and Pet

When it comes to display animals, the death's head cockroach is a great one for the collectors. The true death's head cockroach is about an inch long, and it is a lovely glossy black. What distinguishes this insect from the rest is a skull-shaped amber or beige marking on top of its thorax. This marking is quite clear on some insects, and more blurred on others, but it makes for a beautiful choice for a pet.

The true death's head cockroach is often confused with the hybrid death's head cockroach. The hybrid version of this species was crossed with a significantly larger cockroach. The hybrid has an impressive size of up to 3 inches in length, but the markings are less clear. Most pet stores do not distinguish between the two, and they are both frequently sold under the same name. The hybrid is by far the more common insect in cockroach keeping circles.

When you are thinking about adding the death's head cockroach to your collection, there are a few things that you need to be aware of. Firstly, this animal can produce a rather sharp, musky aroma when it is distressed. This is something that is harmless, but it is a little unpleasant, and it can be startling if you are not ready for it. This smell dissipates quickly, but those with sensitive noses might want to give this animal a pass.

They are not climbing insects, and thus they will need more surface area at the bottom of a tank if you want to keep a colony. They are fairly docile, and though they are faster than a Madagascar hissing cockroach, they are not as speedy as some other species.

It is also worth remarking that though they have wings, they do not fly. Instead, they prefer to jump or to glide from point to point. If you are holding a death's head cockroach, be aware of the fact that it might leap out of your hands and glide to a place where you cannot see it.

On the whole, however, they are fairly easy cockroaches to keep. They are hardy, and the hybrid version can easily be used for feeders. The true death's head cockroach is a slow breeder, however, and it may take several months before you have a thriving feeder colony. This is a fantastic cockroach to keep as a pet, however, and you can be sure that the distinctive markings will impress anyone who happens to see them.

13) Dubia Cockroaches

Type: Feeder

When you are looking for a great cockroach to use as a food source for your other pets, the dubia cockroach is generally considered the superstar. This cockroach hails from South America, and it is a fairly calm cockroach that is typically found in the wild, rather than in people's homes.

One of the big advantages in keeping dubia cockroaches is that they are sexually dimorphic. That means that it is easy to tell the males apart from the females. This is less of an issue if you are planning to keep cockroaches as pets, but it is invaluable if you are planning to breed them.

The females are about 1.5 inches long and the males are just a little shorter. The females are much bulkier than the males, which are more slender and streamlined. The female also has yellow and black bands across the abdomen, while the male is a chestnut brown.

Female dubia roaches produce between 25 and 30 eggs at a time, and they are very easily bred in a normal container. All you need to do is keep the heat up and the humidity high. Soon enough, you'll see plenty of nymphs in the enclosure.

Another advantage of keeping this cockroach is that it is relatively slow. If you drop it, it will not go zipping across the floor to immediately enter a crack in the wall.

On top of that, because it is a tropical species, it is unlikely to breed even if you have a few escapees. If you keep your home at a moderate temperature, they will not breed at all. On the other hand, if you live in a tropical climate yourself or if you are having a heatwave, they may start to reproduce young in your walls.

When you are looking to keep dubia cockroaches for feeders, you may be best off considering a dry set up. A dry set up has no substrate in it at all; it is just a plastic bin with some climbing items to provide more room, some food and a water source. While this is not the most attractive setup in the world, you will find that it is quite simple when you are looking to harvest insects for a meal for another pet.

A dry enclosure is also much easier to clean, allowing you to simply sweep out the droppings on a regular basis.

14) Kenyan Cockroaches

Type: Feeders

As the name implies, these cockroaches hail from the African country of Kenya. They are generally considered the smallest feeder roaches on the market, and they have a good reputation for being easy breeders and for having simple care requirements. Kenyan cockroaches are less than half an inch in length, and they do best when they have a substrate that is in a chip-like format, like wood or pellets. They are simple to sex, as the males have fairly short wings on their bodies, while the females have wings so small that they are hard to see. Males tend to climb more than females do.

One interesting trait of the Kenyan cockroach is that the female carries her egg sac inside her body, resulting in living young. Kenyan cockroaches also love to burrow. The females will create birthing chambers before they start to produce their young.

While it is generally a good idea to keep cockroach species apart, Kenyan cockroaches have anecdotally been kept with larger species. While this does reduce the time required for care, it is also important to remember to provide the Kenyan cockroaches with a place to hide. They need wood chips that they can burrow in to get away from larger cockroaches.

15) Discoid Cockroaches

Type: Feeder

When you are looking for a medium-sized feeder cockroach, one of the most robust options out there is the discoid cockroach. The discoid cockroach has been kept for more than thirty years in the reptile community, and they are known to be easy breeders and very hardy on top of it.

There are several advantages to using discoid cockroaches for feeders. In the first place, they neither fly nor climb. That means that they are not prone to escapes. On top of that, the discoid cockroach is a tropical species. It cannot escape to breed in your home as long as you keep your air conditioner on.

Another reason to pick up the discoid cockroach is that it has a rather docile nature. This means that it will not menace the pet that you try to feed it to. That, combined with its lack of climbing or flight ability, means that even the slowest or youngest reptile should have no problem catching it.

This is an excellent species when you are looking for a healthy feeder colony.

16) Lobster Cockroaches

The lobster cockroach originally hails from the tropical Caribbean. This rather plain looking cockroach grows to be just a little over 1 inch long, and though they can climb, they are not skilled at climbing glass.

What makes this feeder cockroach so notable is the fact that it is one of the most fleshy cockroaches out there. It has a relatively soft-shelled body with plenty of meat on the inside. This is something that suits pets that might be older or simply not so good at hunting and extracting the meat of their prey out of its shell.

They are known to be fairly poor at hiding from predators, and on top of that, they do not even release a bad odor when they are threatened; as some other cockroaches do. They will not bite your pets at all.

This is a relatively short-lived species, which has some advantages when it comes to breeding them. It takes a nymph between 3 to 4 months to mature under optimal growing conditions, and all in all, the insect lives for about a year. That means that it is possible to get an established colony within 6 months, even when you are taking away specimens to be eaten by your pets.

It is worth noting that this species is known for being quite good at climbing. Some care must be taken to prevent it from crawling out of its enclosure. On top of that, it is also fairly speedy. Diligent cockroach keepers should be fine, but if you are worried about escapees, you might need a slower cockroach.

17) Turkistan Cockroaches

Type: Feeder

When you are looking for a cockroach that will be a good breeder and a fast grower, you might want to consider the Turkistan cockroach. These cockroaches are originally from tropical areas, and they are a little less than an inch long when they are fully grown.

This is another cockroach that is fairly speedy, but unlike the lobster cockroach, it does not climb, making them a little less prone to escape. They also do not fly.

The Turkistan cockroach matures at about five months, and the adults live for between half a year to a year. They can only be sexed as adults. Look for wings, as only the males have them.

18) Green Banana Cockroach

Type: Feeder

When you are looking for a smaller feeder cockroach, the green banana cockroach is one that you should definitely consider. This cockroach grows to be only ¾'s of an inch in length, and though the adults can fly and climb, the nymphs do neither. On top of that, you will also discover that these cockroaches are very soft, meaning that they are an excellent choice when you are looking for prey for lizards like chameleons, which have somewhat tender jaws.

Roach Crossing in the United States and The Roach Hut in the UK are both excellent places when you are thinking about finding cockroach breeds that suit you.

Chapter 4) Should I Own Cockroaches?

1) Are Cockroaches the Right Pet for Me?

The question of whether you should have cockroaches is one that you should answer decisively before you purchase them. It can be easy to get carried away, especially if you love insects. There are many factors to consider before you make the decision to take on these interesting and complicated pets.

Firstly, do not get a cockroach simply because you want to scare people. There is a certain revulsion reaction that many people have for cockroaches. After all, they are usually synonymous with unpleasant things like poor housing, unhygienic conditions and poverty. They are not typically associated with good things.

Cockroaches do not deserve the reputation that they have. Like any other creature, they are trying to survive, and through no fault of their own, their survival can often make our living conditions frustrating.

It is never kind or wise to choose a pet simply because it makes other people upset or unhappy.

A closely related question that you need to answer is whether anyone you live with will be unhappy with these pets. Talk to the people around you before you decide to get these animals, and make the proper arrangements. You may find that some roommates or family members have enough issues regarding these animals that it would be unkind to get yourself a colony, even if it is one you kept in a private area.

Some people choose to take on raising roaches because they are afraid of them. Exposure therapy is often used as a way to desensitize a person from a fear, and keeping a colony of roaches is one way to prevent a fear of insects from becoming debilitating. This is a good idea if your fear of insects is relatively minor, but if you have a serious phobia, consult with a therapist before you order a colony of cockroaches online.

You may also choose to keep cockroaches because they are a good source of food for your other pets. Snakes, lizards and even fish can benefit from being fed live insects, and if your pet was wild-caught, that is, not trained to eat anything but live prey, having a colony of live cockroaches on hand can be very handy.

Are you in the educational field? Science teachers who instruct children at all grade levels can benefit from having a few cockroaches around. Children form their impression of what is interesting and what is frightening early on, and acclimatization can make them a lot more comfortable with animals that they may have considered frightening or gross before. They are easy to care for so they will not be a burden in a busy classroom, and they can even be taken home by students who are curious about them over a long weekend.

Cockroaches are also very hardy. Their reputation states that they can live through just about anything, and though this makes them into very easy pets, that does not mean that you should neglect them. They are often considered a good starting pet for children because they require a little bit of responsibility but the not the same amount that would be required for a dog or a cat.

Cockroaches require less time, investment and energy than other animals, but make sure that you choose them as pets for the right reasons.

When you get to know more about these insects, you will see that they are quite amazing in their own right. Perhaps the best reason to get cockroaches is because they are fascinating insects that can teach you a lot.

2) Solo or Group?

Many people wonder how many cockroaches they should keep. As far as may be determined, these insects do not suffer when they are kept alone. A single solitary cockroach can get along just fine when they have enough food and water.

On the other hand, they do not mind being crowded into an enclosure with other members of their species either. Though it varies from species to species, most cockroaches are fairly gregarious. They do not mind the presence of other cockroaches of their own kind, and for the most part, they do not fight.

When you are specifically raising cockroaches as a type of dietary supplement for your other pets, you will find that it is a good idea to start with quite a few. That way, you can encourage some to breed while making sure that your pet still gets its meals.

However, given the fact that many people start off with nymphs, it can take several months before you are reliably feeding out of your cockroach colony. In many cases, starting off with a colony of twenty or thirty cockroaches is usually a good idea. Some people start off with more than that because they want to make sure that they can feed their animals even as they get their colony established.

Another thing to consider is rental properties. If you do not own your own home, you may have a landlord who is leery about allowing you to have pets that are commonly known to be a nuisance associated with poor properties. Many people are tempted to hide their pet cockroaches from their landlords, but this does carry some risks.

For example, if it is discovered that you are breaking your lease, this may be grounds for eviction. On the other hand, even if your cockroaches are hidden, they may be prey to pesticides that the landlord brings in.

Across the board, it is important to be truthful to the people from whom you rent. If you rent from a single person rather than from a large company, you may be able to petition for the cockroaches if you can show how you keep them, how secure they are and how little a chance of escape there is.

If you want to start small, simply consider getting one or two cockroaches. They will not suffer from being kept in such small groups, and you can decide if you really want to take on this type of pet.

Cockroaches kept singly do seem to live longer than ones kept in colonies, but you may also want to make sure that you have a colony on hand because it is simply entertaining to see cockroaches interact as they would if they were in the wild.

3) Where Do I Get Cockroaches?

There are several places to get cockroaches. If you have a good pet store in town that caters to the exotic pet trade, you may see them there either as a supplemental food for other animals or as pets themselves. If you are looking for cockroaches at the pet store, you will largely see hissers of one sort or another there.

If you are looking for more exotic cockroaches or you do not have an accommodating pet store in the area, you will discover that it is usually a good idea to search online. There are a number of breeders who will sell you cockroaches from their site, but before you jump right in with your credit card, there are a few things to keep in mind.

First, you are typically going to be able to get very young cockroaches. They are cheaper, and they are very simple to raise. If you are creating a feeder colony, you will find that it is best to purchase between twenty and thirty young cockroaches and to feed them until they are breeding and providing enough young.

It is worth being a little wary of anyone who is selling a large number of adult cockroaches for a very small amount of money. This may seem like a great bargain, especially given that there is a percentage of young cockroaches that simply do not thrive, but this is a situation where you get what you pay for.

There are unscrupulous dealers out there who are simply providing spent breeders when they are selling full adults. This means that these are older insects that are at the end of their life cycles.

Buying spent adults is a perfectly reasonable thing if you just want to feed your animals, but if you want to breed them yourself or you want to keep them as pets, it is usually worth buying younger animals.

When you are buying cockroaches in person, take a good look at them. The cockroaches should have all of their own legs, and their antennae should be whole. If you see them during the day, you may notice that they are relatively sluggish. There is in fact nothing wrong with this, as cockroaches are mostly nocturnal. They do not start to get really active until it is dark.

If you have never kept cockroaches before, it is good to start small. However, it is also worth noting that it is not noticeably harder to care for a dozen cockroaches than it is to care for two. Cockroaches do not eat a lot, and as long as you have the space to keep them, you'll be feeding them all at once anyway.

4) Mixing Cockroaches

Many people wonder if you can mix different species of cockroaches. The answer is typically yes, but there are some caveats.

First, only mix cockroaches that have similar housing and feeding requirements. If they have different needs, house them in different containers.

Second, only house cockroaches together that are of a similar size or, as in the case of the lobster cockroach, the smaller species is known for cohabiting well with larger species. If you house a large species with a smaller species, the larger cockroaches might eat the smaller ones. On top of that, even if this type of aggression is not an issue, larger cockroaches can prevent smaller ones from getting to the food.

Some people also wonder about housing males with females. There are a few things to consider before you try a mixed sex group.

First, if you house males with females, you will get eggs and eventually more cockroaches. Cockroaches breed easily enough that this is a given as long as you care for them relatively well.

There is also evidence that states that for the most part, male cockroaches are more aggressive than females. With most of the cockroaches that people keep as pets or as feeders, it is very easy to tell the difference between the males and females, so it might be worth your while to keep an eye on the population.

Some people much prefer to keep more females than males; not only does it reduce the amount of aggression that might be occurring in the colony, you will also discover that it increases the amount of young that are produced.

5) Lifespan

It seems that the larger the cockroach is, generally the longer it lives. The Madagascar hissing cockroach, as well as most hissers, lives anywhere from two to five years. Most smaller cockroaches simply live between one to two years, and the rhinoceros cockroach can live to be 10 years old.

It is worth noting that cockroaches that are kept in smaller groups tend to live longer. This is preferable if you are keeping your insects as pets. When they are kept alone, the females are not worn out from producing young on a regular basis, and the males are not constantly stressed due to dominance displays.

Consider how long you want to keep your cockroaches and think about what that might mean to your husbandry choices.

6) Cockroaches and Children

Whether your child loves insects of all types or you are simply concerned about keeping cockroaches in a house with children, there are a few things to consider.

If you are thinking about getting cockroaches for your children to keep as pets themselves, you cannot do better than to pick up a pair of Madagascar hissing cockroaches. These large insects are known for their docility and their ease of care. They do not bite, they do not produce an odor, and they are fairly resilient. Madagascar hissing cockroaches are hardy enough that they will not suffer from a child's occasional lapse in memory, and they are large enough that even a rather clumsy child can be taught to handle them without harming them.

It is worth mentioning, however, that these cockroaches have fairly sharp spikes on their legs. They do not use these spikes defensively, but someone who grabs a cockroach of this species with their hand in a careless way might cut themselves. If you want to give your children cockroaches for pets, you should always teach your child how to hold them.

If you are largely rearing your cockroaches as feeder animals rather than as pets, there is nothing wrong with keeping them in the same house with children. Some cockroaches produce a slightly musty odor, and some will bite if provoked, but for the most part, they can be left alone.

If you have curious children and a thriving cockroach colony, some base rules should be in place. Your young children should not open the lid or try to remove any insects without your permission, and if you have a rather squeamish child, you may wish to simply house the insects in an opaque plastic container rather than in a terrarium.

Cockroaches that are bred as pets or as feeders are very safe for children to handle, though it might be worth teaching your child the difference between these animals and the ones they might see from time to time in the wild.

7) Cockroaches and Safety

When you are thinking about acquiring cockroaches as pets, it is important to remember that the cockroaches that you see in the wild or hiding in the home are different from the ones that you will keep yourself.

In the first place, you are likely to be looking at different species, and in the second place, pet and feeder cockroaches are not harmful to people. Wild cockroaches can carry diseases and toxins that they acquire when they crawl through homes. This is something that makes them downright dangerous to have in or around your food source or food preparation area. On top of that, several wild cockroaches can bite quite hard.

When you are looking at the popular cockroach pet species, the first thing to remember is that they are bred and kept in sterile conditions. They have never gone anywhere where they would pick up something dangerous, and the things that they eat have been specially prepared for them.

On to top of that, most of the species kept as pets do not bite at all. The ones that can bite only give you a sharp nip.

However, it is worth noting that you should always wash your hands after handling your cockroaches. This prevents the spread of bacteria, and it is simply a good husbandry practice in general. This is especially important if you have isolated one group of cockroaches for a health issue and do not want to infect the rest of the colony.

Also remember that if you are feeding your cockroaches to reptiles, you should always be careful to wash your hands after feeding your cockroaches to the reptiles as well. Reptiles carry salmonella, and if you are not careful, you will find yourself spreading salmonella throughout your home after a straightforward feeding.

Cockroaches are an entirely safe pet to keep as long as you know how to take good care of them and how to maintain hygienic procedures.

8) Licensing Issues

When you are thinking about keeping cockroaches, you must first figure out whether the cockroaches that you want to keep are legal in your area. Every state has different rules, but for the most part, the statutes state that you can keep foreign insects but not local ones. In environments where foreign insects can become rampant, however, there are usually further rules that are imposed to prevent the hazard to the local environment.

For example, the Madagascar hissing cockroach, which is the species most frequently kept as a pet, is not native to the United

States, but it is under certain bans in the United States. In the state of Florida the law states that the only hissing cockroaches that may be kept by pet owners is the male. This is designed to prevent possible escapes and subsequent infestations of local areas. Further, Florida also prohibits shipping in exotic cockroaches.

On top of that, due to stringent licensing issues, there are some states where it is legal to own cockroaches, but illegal to sell them. This may mean that you have to do some traveling to get the cockroaches that you want.

Before you purchase any pet, make sure that it is legal where you are. If you are thinking about getting cockroaches but you are unsure about whether they are legal for your area, the best place to figure it out is with your state's Department of Natural Resources. They can either answer the question for you or direct you to a person who can tell you what the truth is.

While it might seem very easy to smuggle cockroaches into your home state, do not do this. There are some severe penalties in place for people who bring in exotic or harmful species, and a stiff fine might just be the beginning of it. If exotic cockroaches are not on the table, let it go and pick another pet.

9) A Word on Veterinary Care

Cockroaches are considered an exotic pet. Not many people have them, and very few veterinarians have any experience with them at all. If you are worried about care for your pets, you will be on your own for the most part, though having the Internet around is a helpful thing.

Instead of talking to veterinarians, consider speaking to entomologists from your local university. Most entomologists are more than willing to help out members of the community with their questions, and they can refer you to people who specialize in cockroaches.

Remember that there will be a certain amount of death in any large colony. The larger a colony is, the more likely it is that you will have deaths that result from starvation or simply due to having some animals that just refuse to eat. In many cases, cockroaches die because they are turned over on their backs. Cockroaches are unable to right themselves, and thus simply die as they are. What this means is that a certain number of deaths are to be expected. You only need to start worrying when the entire colony looks like it is being affected.

10) Training Cockroaches

We see cockroaches in the movies and on television, and these roaches are trained to run from one spot to another. However, it is important to remember that cockroaches are not necessarily intelligent animals and that they operate primarily on instinct. They are never going to be as skilled or perceptive as an animal like a dog or a cat, and there are certainly limits to what they can do.

It is also worth noting that some kinds of training are possible. For example, it is very clear that cockroaches have an extraordinary sense of smell, and they can easily tell the difference between various scents.

In a recent behavioral experiment, scientists taught cockroaches to run towards the scent of peppermint rather than vanilla. On their own, cockroaches prefer the scent of vanilla, but the scientists associated the peppermint with sugar water and the vanilla with saline. In very short order, the cockroaches quickly realized that pursuing the peppermint scent would result in rewards.

If you want to try a fun game with your cockroach, you may try to build it a maze out of wooden blocks with a reward of sugar water at the end. To begin with, you might want to create an association between a treat like sugar water with a particular scent, and then you can bait the maze.

Interestingly enough, this would be the exact procedure that would be used for training cockroaches as first responders. Cockroaches are very durable and highly suited to climbing through rubble. In the event of accidents or earthquakes, a cockroach equipped with a camera and radio device could be sent through the rubble to find trapped survivors. This type of technology is still being developed, but the results could save lives.

Some people have managed to get flying roaches to come to their hands by persistently holding sugary treats in their hand, but this is far from reliable.

Across the board, the main activities for those with pet cockroaches is observing them and holding them. They cannot be trained to do tricks the same way that most mammal pets can be trained, and on the whole, they are happier to be left alone than they would be if they were being handled all the time.

11) Cockroaches Versus Crickets

Cockroaches are excellent feeder animals, but they are far less popular than crickets. The truth of the matter, however, is that when these two animals are judged objectively, cockroaches are by far the better insect to feed to your animals.

The main advantage of crickets is that they are common and that they are cheap. You can walk into just about any pet store and find crickets for sale. They are typically sold in the hundreds, and they are quite cheap on top of that.

Cockroaches are a little more obscure and a bit more expensive, but they are hands down the better choice.

First of all, cockroaches live longer that crickets. Crickets mature quickly, but they die within two months. That means that it takes a lot more time and effort to maintain a cricket colony, where there will always be more animals dying than being eaten. Compare this to cockroaches, which can live several years and produce clutches of young the entire time. Once you have a breeding colony of cockroaches, your food needs are met.

Another thing to keep in mind is that crickets smell bad. They produce a musty smell that is quite distinctive, and it is one that can quickly soak into a room. Some cockroach species produce a strong smell when they are frightened or when they want to get away from predators, but on the whole, they have absolutely no smell at all.

On top of that, crickets are jumpers and need to be kept carefully or you will risk escapes. Cockroaches are extremely varied from species to species, and there are some that are noted for their docility, their slow speed and their lack of interest in escape. Do your research and if escapes are a concern, look for a wingless cockroach that does not jump or climb. Crickets are far more active than cockroaches, meaning that they can be difficult for your pet to chase down and eat. This is especially a problem when you are dealing with older animals. While there are some cockroach species that are known to be quite speedy, there are also others that are quite slow. They make easy prey for your pet and they do not offer a challenge at all.

Perhaps the greatest advantage that cockroaches have over crickets is the fact that there are so many varieties. Whatever type of animal you are trying to feed, there is a cockroach that is designed to handle its issues. There are cockroaches that have softer shells, which is better for animals with weak teeth or jaws, and there are cockroaches that are much slower and poorer at hiding, to make up for animals that do not know how to hunt.

Cockroaches can establish breeding colonies in six months or less, and at that point, all you need to do is feed them. After that, you no longer need to spend money on this type of live food for your animals. Crickets, on the other hand, can be made into breeding colonies, but they are more difficult to control. It is all too likely that you will simply have too many or one day find that most of your colony is dead!

Cockroaches are worth the time that you put into them, and after the initial investment, which is pricier that what is required for crickets, the savings can be impressive.

12) Handling Your Cockroaches

Some people choose to simply keep cockroaches as an observational pet, with very little interaction, but cockroaches of all sorts can be handled safely and easily. You will also need to handle the cockroaches if you are planning to feed them to other animals, so knowing how to hold them properly is handy.

First, the bigger the cockroach is, the easier it will be to hold. You will also find that some cockroaches are simply more docile than others. By far the easiest cockroach to handle is the Madagascar hissing cockroach. It's large size, docile nature and wingless state make it ideal for casual handling.

To make sure that you are not harming the cockroach when you pick it up, grasp it firmly between thumb and finger. Only grab your cockroach at the thorax, which is the hard segment directly behind its head. Do not grab the cockroach by its head, its legs or its abdomen, which is the large part at its rear.

When you grasp your cockroach by the thorax, do not pull it away from where it is right away. Move slowly, and do not yank. The issue is that the pads on the cockroach's feet allow it to cling to the surface that it is on. If you pull too fast, you risk damaging the cockroach. In a worst case scenario, one or more limbs might actually pull of.

The best way to spend time with your cockroach is to simply let it run through your hands. These insects will generally not be still on you, so you must pay attention to it while it is on you. There is a slightly ticklish sensation as they walk on you. To encourage a cockroach to slow down, stay very still and make sure that you are not sitting or holding the cockroach in a strong light.

If you are interested in making sure that your cockroaches grow more tame, it is worth your while to handle them a little bit every day. Do not stress out these animals too much by handling them too often, but spending ten minutes or so every day can help your cockroach learn that you are not something to be afraid of.

Sometimes, you can even encourage your cockroaches to look forward to you lifting them up. Swiping a little bit of sugar water into the palm of your hand before you lift them up can make them much happier to see you.

13) Getting Rid of Cockroaches

There may come a time when you can no longer care for your cockroaches. If that time comes, under no circumstances should you release them into the wild. Not only is this an irresponsible thing to do, you will also find that it can have a disastrous impact on your local ecosystem.

Cockroaches are nothing if not tough, and they can thrive in a new environment that is not suited to them. In many cases, they may out-compete the local insects and cause problems in the food chain.

If you need to get rid of your cockroaches, ask around to see if anyone wants them. If they are a large and docile species, it is worth offering them to your local school. The science department might enjoy having a few new specimens to work with.

You may offer your cockroaches to the zoo. They are usually willing to take donations of animals, particularly if it is a variety that they do not have yet.

If you are trying to get rid of a feeder colony, advertise online and see if any of the local population of reptile-keepers wants them. You can charge for the colony or you can simply let them go for free if you are in a hurry.

It may also be worth your while to bring your cockroaches to your local pet store. In some cases, the pet store will buy them for you, or they can put you in touch with someone who is willing to purchase them.

There are avenues where you can get rid of cockroaches, so do not simply let them go!

Chapter 5) Care and Keeping

Like any pet, cockroaches require good housing, good food and competent care. Consider making sure that you know more about what your cockroach needs before you start purchasing specimens that appeal to you.

1) Set Up and Monthly Cost

Overall, the cost of keeping cockroaches is very low. Depending on the route that you choose to take, feeder cockroaches can cost as little as five dollars for 20 feeder nymphs, with large ornamental pet cockroaches costing between three and five dollars depending on where you get them. In the UK, the prices seem to be a little steeper, at 10.75 pounds for 20 feeders and 6.50 pounds for some of the more exotic cockroaches.

The initial outlay of expenditures is going to be the most your cockroaches will ever cost you. A ten gallon aquarium can cost you about 25 to 35 dollars in the US, while costing about 40 pounds in the UK. However, aquariums are relatively easy to come by used, so look around online. A heat pad costs about seven dollars in the US, and about 10 pounds in the UK.

You may choose to spend five dollars or three pounds on substrate every few months, or if you decide on a dry enclosure, you can skip it entirely.

For food and water dishes, you can simply use old dishes that you have in the home; there is no need to purchase anything special for it.

When it comes to food, cockroaches are amazingly easy to feed. The dry dog kibble will be the most expensive part of the cockroach's food, costing about 14 pounds for 17 kilograms in the UK and about 40 dollars for 40 pounds in the US. This can come out to less than a dollar or a pound per month if you shop carefully. Simply use fruits and vegetables from your own meals, or budget about 10 dollars or about 6 pounds total, and that is your cockroach's budget.

2) Enclosure

When you are looking to make your cockroaches feel at home, the three words to remember are "dark," "warm," and "damp." Most cockroaches are originally insects that live in rotting wood or in the darker parts of human habitations, and while they do not require anything fancy, you will discover that there are certain needs that should be considered.

First, think about the space that is required. Cockroaches are gregarious, and if you are keeping a colony of around ten or fifteen or so, you will discover that a ten gallon aquarium is a good choice. One of the nice things about purchasing an aquarium for your cockroaches is that you do not have to worry if they are water tight at all. You can even seal up obvious cracks with tape if you are not all that concerned about appearances.

If you are keeping just one or two cockroaches, you can often get by with a light plastic container with a ventilated, snap-on lid. These single gallon containers are often used as transport options for small pets, but they make an ideal permanent home for just a few cockroaches.

Cockroaches have pads on their feet that allow them to cling to just about any surface, and that means that you cannot expect them to stay in the enclosure unless you take certain precautions. First, your container needs a lid. The lid should fit closely to the container, and ideally it should snap on. However, given the fact that cockroaches can flatten themselves very easily, this is not the only measure that you should be taking.

On a regular basis, make sure that you smear Vaseline or olive oil around the top three inches of the aquarium. Use a piece of cloth to coat the slippery stuff completely around the interior edge of the enclosure.

When you make it this slippery, you can prevent the cockroaches from getting out. Some people only use the oil or Vaseline, but this does not help when you are dealing with insects that fly.

If you are breeding cockroaches for feeder or scientific purposes and you want to make sure that you have plenty of room, there are other alternatives. If you are not interested in display and you need a large number of animals, you may decide to put them all in a large plastic garbage can with a ventilated lid. This gives you plenty of space to work with.

Another enclosure option that rests between aquarium and garbage can is that of a plastic bin with a snap-shut lid. Some people also cut large holes out of the top and cover them with mesh for better ventilation, allowing you to see the cockroaches as well.

Your enclosure should have a dish for feeding your cockroaches, a dish for a water source, a substrate and a heat source, all of which are discussed below.

The enclosure you choose should be one that suits your home and your needs for your cockroaches. When you are thinking about how you can make sure that you have the right one, be willing to do some research and to look at your own home. Figure out how you can make sure that your enclosure is the right one.

There are plenty of places to get great enclosures. Check PetSmart and Petco in the US and Swell UK in the UK.

3) Lighting

Cockroaches do not need light to thrive. In most cases, cockroaches are most active in the dark and light typically means that they are going to scuttle away and hide.

Do not leave your cockroach enclosure in the way of a window, and do not turn a lamp on it. This can overheat the cockroaches in the enclosure, and it can make them nervous if there is no way to get away from the light.

Instead, make sure that the lights that you use around them are relatively dim. This means that when you turn them on, you will not immediately panic your cockroaches and send them scuttling for cover.

4) Creating Space

Cockroaches will happily climb on any surface that you care to name, but the truth of the matter is that they do have preferences. If you have a glass terrarium, you will discover that some cockroach species will climb on the glass, but they will appreciate other options as well.

While cockroaches do not mind being in crowded spaces, they will certainly take the opportunity to climb. One excellent choice to bring a little bit of height to your cockroaches' enclosure is to use egg cartons. Paper egg cartons have a fair amount of texture, making it easy for cockroaches to climb up, and you will also discover that some cockroaches will happily snack on the cartons themselves. Paper cartons are also easy to recycle, and they hold water very easily, meaning that they can help hold humidity in the tank. However, you must be aware that they can also rot, and thus must be replaced on a regular basis.

If you want to place paper cartons in your enclosure, simply throw them into the bottom of the unit, stacking them loosely on top of one another. The crevices created help your insects feel more at home in the dark in the tight spaces.

You can also add height to the tank by throwing in branches and plastic toys. Some people love to create an entire set for their pets, allowing them to rampage through model city streets or over resin models of gorgeous architectural features.

Dead branches are a great choice for your cockroaches' enclosure, as they are what roaches would climb in the wild. However, if you do decide to get your cockroaches branches for their enclosure, make sure that you buy them from the pet store rather than dragging them in from the wild. Branches that you find outside can have mites and parasites in them that can be a problem.

If you do want to bring branches from outside, make sure that you pull off all the bark, which can hide and protect insects, molds and fungi. At that point, freeze the wood inside a sealed plastic bag for a week. This will get rid of most of the issues, but to get rid of all of them, you will need to think about boiling the wood.

Experts caution against boiling wood because it could have air pockets that overheat and explode. However, many people find that they can do just that very easily and safely. You should be aware of the risks before you proceed.

To boil your branches, simply bring a pot of water to boil and drop the branches in. Boil the wood for half an hour, filling up the pot as necessary and then allow it to dry thoroughly. After you do that, you will discover that it is very simple for you to place the branches in the tank without fear of parasites.

Of course branches from the pet store are best. These have usually been stripped of any problematic fungi, molds or insects in a kiln, and they even have stands that allow you to place them where you want them.

Some people choose to go the extra mile and place real plants into their enclosures. While this is quite lovely, it is usually a poor idea unless you are ready, willing and able to keep everything in good order. Cockroaches tend to gravitate towards rotting material, but they will generally eat everything that they can find. They are more than willing to help themselves when it comes to your nice plants, so it might be best to stick with plastic.

5) Substrate

The substrate is the substance that you use to line the bottom of the tank. Scientists who are studying cockroaches will usually bed them on substrate of some sort. However, there are people who breed cockroaches who state that a substrate is not necessary at all, and it is just something that needs to be cleaned. Dry enclosures are enclosures where there is no substrate at all. They are typically used by people who keep cockroaches as feeders and do not intend to display these insects or interact with them.

If you are planning to display your insects, on the other hand, substrates add a certain aesthetic value to your enclosure. You may also wish to think about aesthetics. Substrates take away from the harsh edges of the enclosure, creating an environment that is more akin to what cockroaches would enjoy in the wild.

Cockroaches do interact with the substrate, burrowing in it and rooting in it as they do in the wild. Think about what you want to see in your cockroaches' terrarium, but also consider what you want to clean.

When you are looking for the right substrate for your cockroaches, there are a few great options available. When you are looking for something that is relatively sturdy, think about throwing in wood chips. Avoid wood chips that are made from aromatic woods like pine or cedar because they can create fumes that give your cockroaches difficulties. Wood chips are great hiding places for smaller cockroaches.

The hardware or gardening store is a good place to start when you are looking for substrates. Both peat moss and sphagnum moss are excellent substrates. They can be used to line the bottom of the tank, and they are very similar to the type of material that cockroaches would be attracted to in the wild. Some people also lightly spray them with water to give their cockroaches another source of water and humidity.

Another excellent substrate is newspaper, though it certainly is not as attractive as moss or wood chips. It is, however, cheaper, and you can get plenty of newspaper for free. If you don't mind spending a little more cash, think about simply purchasing pellets made from newspaper or shredded recycled paper to use as your cockroaches' bedding.

When you have brought the substrate home, give it a quick once over to figure out if there are any mites in it. This is something that can save you a lot of trouble in the long run. Mites are tiny insects that can latch on to your cockroaches, and once you have them, they can be difficult to get off.

Look for very small, very pale insects that move relatively quickly. If you see any movement in your substrate at all, return it to the place where you bought it and ask for a refund or a replacement. Mites can happen in nearly every medium, so look closely.

6) Heat

Most cockroaches need warmth if they are going to thrive. Madagascar hissing cockroaches especially require warmth because they originally hail from a tropical climate. If the temperature drops too low, you will find that your cockroaches get slower and more sluggish. If the temperature drops low enough, it might even kill them.

There is an acceptable heat range for your cockroaches, and this range holds true for all of the species that most people keep for pets or for breeding. The ideal range for an enclosure of cockroaches is between 72 degrees Fahrenheit and 80 degrees Fahrenheit. This is what it takes for your cockroaches to stay healthy. The closer to the top of the range you get, the better off your cockroaches generally are. As a matter of fact, tropical species like the Madagascar hissing cockroach live in environments that regularly climb to more than 100 degrees Fahrenheit.

You should make sure that the temperatures do not drop too low. While cockroaches can frequently survive temperatures that are 60 degrees and lower, the issue is that they get slower and more sluggish. Cooler temperatures will also make it more difficult to breed your cockroaches.

There are several ways to make sure that your cockroaches' enclosure stays warm enough. Choose the way that best suits your home and your lifestyle.

First, you may choose to simply keep the heat in the room in question at the right temperature range. If you live in places where it is warm for most of the year, this is not really a problem at all. However, this can be an issue if you are in a place where the temperature drops. Some serious cockroach keepers simply keep a small dedicated room warm enough for their cockroaches, but this can be a bit of a burden if heating costs go high in your area.

One way to keep the heat in a cockroach enclosure high is to make sure that you keep a space heater or a ceramic heater on throughout the day. This is something that can make a huge difference to the temperature without raising the heating bill too much. However, you should definitely keep in mind the fact that some of these heating devices should not be used unless there is someone else in the room with them.

Some people choose to heat their cockroaches using a heat rock. Heat rocks are sold at many reptile and pet stores, but they are not ideal because there is no way to regulate the heat coming off of a heat rock. These heat rocks can overheat, and in doing so, they can seriously injure or kill a cockroach.

The best way to heat a cockroach enclosure is to simply get a heating pad and place it underneath the tank. A good heating pad has settings, and by placing it under the tank, you do not allow the roaches direct access to it. This is something that protects them from possible burns.

If you live in a naturally warm environment, some people advocate keeping cockroaches outdoors. This is something that has both advantages and disadvantages. Cockroaches are very hardy, and if you live in a space that is similar to their natural environment, they will probably thrive.

However, keeping the tank outdoors does not give you much control over what they experience. A cold snap can destroy your breeding efforts, and wild animals can find their way to the tank and chew it open leading to many escapees. Cockroaches that escape from an outdoor enclosure may enter the environment, harming the local ecosystem.

Though there are some advantages to keeping cockroaches outside, it is largely considered a better idea to keep them indoors in an environment that you alone control.

7) Cleaning

Despite their association with filth and disease, cockroaches are animals that produce very little scent on their own. Cleaning their tank is a relatively straightforward procedure, and it only needs to be done once every two to three weeks, depending on the size of the enclosure and the number of cockroaches that you are keeping. Some cockroach keepers go for even longer, especially if there are only a small number of animals in the tank. Frass is the term for cockroach droppings, and it is very dry and odorless. When you are dealing with a dry enclosure, it is often enough to dump most of the adults into a clean plastic bin, to sweep up the frass and then to remove the nymphs from the frass using the method described below. However, cleaning an enclosure with substrate takes some time.

First, start by removing everything that the cockroaches climb in or on. Brush the cockroaches back into the enclosure as you do so. For instance, if you keep cockroaches in paper egg cartons, give the egg cartons a sharp tap against the side of the tank to make them drop off.

Remove the food dish and the water dish as well. This is a good time to give them a quick scrub.

Shove everything to one side of the tank, including substrate, cockroaches and detritus.

Quickly place a bowl of fresh food on one side of the tank, which brings the adults over to the clean side. As they are eating, sweep the substrate into a resealable plastic bag.

If you are breeding cockroaches, the substrate and the frass is likely to be full of nymphs. To get the nymphs from the substrate back into the enclosure, throw an empty toilet paper tube into the bag and leave it alone for ten minutes. When you come back, many nymphs will be inside of it. Knock the nymphs back into the enclosure and repeat until the substrate is nymph-free.

It is worth noting that if you are breeding a feeder colony, you are probably going to find dead adults in the substrate. Cockroaches grow old and die just like any other animal, and a small amount of die-off is normal.

When you are done cleaning the enclosure, remember to reapply the coating of Vaseline around the lip of the container. During the course of cleaning, it can get smeared away, leading to your cockroaches escaping.

Do make sure that you tie the bag full of frass and old substrate closed as well. This prevents any stragglers from escaping.

Depending on the size of the colony and the maturity of the colony, you may not have to clean all that often. The truth is that cockroaches can survive through intense neglect, but do not take advantage of this fact.

Interestingly enough, the frass is densely packed with healthy elements that work well for a garden. If you are certain that the frass has no nymphs in it, you can even add it to your garden's compost to enrich the soil. However, you must be certain that there are no escapees in the frass, or you may have introduced a serious pest to the environment.

8) Water Needs

Cockroaches need water just like every other animal, but you must make sure that they do not drown in it. These are not necessarily intelligent animals. They operate largely on instinct, using the organs in their head and on their antenna to navigate the world. What works well in the wild does not always work well when they are living in a man-made enclosure.

When you are creating the right kind of enclosure for your cockroaches, remember that you should have a shallow dish for your cockroaches' water needs. However, this dish should not be filled with water. No matter how shallow it is, there is a chance that your cockroaches might climb into the water and end up drowning. This is especially true if you have small cockroaches or if your cockroaches have bred recently and there are many nymphs in the enclosure.

Instead of filling the dish up with water, simply soak a few cotton balls or a small sponge with water. Place the wet sponge into the dish and place it in the enclosure. Keep the sponge moist, but do not allow it to grow mold or to spread mold to the rest of the enclosure. Every day, make sure that you rinse the sponge out and reload it with fresh water.

The cockroaches will find the sponge and drink water from it. This is very similar to how cockroaches find water in the wild, and they will take to it without a problem.

Do not allow the sponge to run dry. While it takes cockroaches a much, much longer time to die of thirst than other types of animals, it can happen.

Another method for watering your cockroaches is called the wick method. The wick method is used by many universities and research organizations when it comes to keeping cockroaches for both breeding and study purposes.

To utilize the wick method, you need a jar, a nail, a hammer, a pair of scissors, and a length of cotton cord. Use the nail and the hammer to punch a single hole in the lid of the jar. The hole should be small enough that the cotton cord slides through it, and not so large that it falls through.

Using a pair of scissors, cut the cotton cord to a length that is equal to the height of the jar including the lid plus 1 inch. Fill the jar with a few inches of water and slide the cord through the hole in the jar. Screw the jar lid onto the jar and adjust the cord so that one end touches the water and the other end pokes above the lid by about an inch.

The water will be absorbed by the cotton wick and travel up the cord, making the end above the jar lid damp. Your cockroaches can climb up on the jar and acquire the moisture through the cotton fibers. You may choose to shred the wick above the jar a bit, giving them a little more surface area from which to drink.

You may also choose to feed your cockroaches with water gel, which is a common product sold in pet stores and online. Water gel is essentially water that is bonded to a polymer. It does not melt, and because it is mostly solid rather than a liquid, it will not drown your smaller cockroaches.

While it is a very clever solution to the drowning roaches

problem, you will find that this is perhaps the most expensive watering method. If you want to cut the costs, purchase the gel crystals and then hydrate them yourself. This cuts back on the price while preserving the benefits of the water gel.

Make sure that you scrub the water dish on a regular basis, no matter which water delivery method you choose to use. Any kind of moisture combined with the living conditions of your cockroaches can create a situation where mildew is a real problem, so be willing to spend some time keeping the water dish clean.

No matter what water solution you choose for your cockroaches' tank, remember that you should always keep an eye on it. Cockroaches need fresh water every day, and if they have to go with out, some serious issues, including aggression and poor shedding can occur.

9) Ventilation

When it comes to cockroach care, across the board, the more ventilation you have, the better. A single cockroach might not need a lot of circulation, but when you are working with a small colony of twenty insects or even a large one with more than a hundred specimens, the air issues can come to a head very quickly.

If you notice that you have cockroaches dying off, you may be in a situation where there is simply not enough ventilation coming through the enclosure. Seeing to more airflow is the first thing you should do if you have an unexpected and unexplained number of deaths.

If you keep your insects in a plastic tub, there are two ways to improve the circulation. Firstly, consider simply punching more holes into the lid. Make sure that you are making the holes from the outside to the inside; otherwise, some small nymphs can make it out of the box.

If you want to be extra sure that the cockroaches that you are raising do not get out, you might want to make a series of small slits rather than holes. If the slits are narrow enough, you will find that this can prevent the cockroaches from crawling out at all. Do make sure that you make plenty of holes, however. Otherwise, you might find that your cockroaches are still having issues with air flow.

A slightly more attractive way to improve the air circulation in a plastic tub is to cover the top with mesh. Simply cut a large rectangle out of the top of the lid using a heavy knife or a heavy duty pair of shears. Then cut a piece of mesh that is slightly larger than the hole and tape it in place securely with duct tape or some other strong adhesive. If you are keeping your cockroaches in a wooden box of some sort, you can even use a staple gun to secure the mesh. This allows you to look into the box, and it also improves the air circulation greatly.

10) Feeding

Cockroaches, like any other animal, need to eat. When you are thinking about what to feed your cockroaches, you will find that you have acquired animals that are very simple and fairly cheap to keep alive as far as food goes.

The general rule of thumb to keep in mind is that cockroaches require about 40 percent protein and 60 percent fruits and vegetables. This keeps a colony thriving, active and breeding, though of course you can vary it.

When you are looking for the cheapest reliable food options out there, check out cat and dog food. Dry kibble has enough protein and nutrients to keep your cockroaches going, and they need very little food at one time; one bag can easily last months depending on what size bag you might have.

A quarter cup of dry kibble can keep ten or twenty roaches happy for several days or up to a week, depending on how big they are. You do not need to vary the diet. Cockroaches are insects, and they do not need variety in the same way that mammals do. Many people feed their cockroaches once a week.

There are plenty of charts out there for how much to feed your cockroaches, but the key is to simply go with what is good for your colony. Try feeding on a weekly basis, and vary the amount. If you only have a few cockroaches to worry about, feeding is quite easy. If you have a large colony, simply be willing to feed your cockroaches a large amount of food and then to clean it out the next day.

You should also supplement this diet with things like slices of fruits and vegetables. These items are important because they supplement the cockroaches' ability to take in water. Many cockroaches in the wild only get their water through the consumption of rotting fruits and vegetables. Some great options for your cockroaches' diet include things like orange slices, apple slices, cucumber slices, slices of bell pepper and pineapple chunks.

Many people who own cockroaches like to experiment with what their cockroaches want to eat. Though cockroaches are definitely opportunistic feeders who will eat just about anything that they can get to, some keepers have noticed that there are quite a few treats that these cockroaches just seem to prefer across the board. They seem to love sweet things, but some keepers have noted a preference for dark green leafy vegetables too, like spinach and kale. Given the fact that these vegetables are quite rich in iron and nutrients, it makes sense that the cockroaches would aim for them.

Don't be afraid to try your cockroaches on something new. Cockroaches are very tough, and they are not picky. If you are curious about giving them some new food, simply toss it into the cage to see what they make of it. Unlike a dog or a cat, which might have issues with foods that are unexpectedly toxic, you will discover that cockroaches can eat virtually anything that a person does.

Remember that food should be removed from the tank before it starts to mold. Mold will not damage the insects, but it can make the environment smell foul, an issue for the human caretakers. Mold and rotting food can also create a situation where bacteria breeds freely. Throw away any fruit that has sat for more than a day or two so that mold does not become an issue.

In the wild, cockroaches are opportunistic feeders, and this means that they are very willing to eat anything that they come across. Their niche in the ecosystem is naturally designated as that of a decomposer. They can take just about any food scrap and reduce it back into compost, though they certainly do not do it as efficiently as nightcrawlers and redworms.

You can experiment with feeding your cockroaches scraps from your table if you choose to do so. Carefully gather up bits of vegetable and grain dishes and place them in the cockroaches' feeding dish. This is a good secondary use for food that would otherwise go to waste.

If you choose to feed your cockroaches table scraps, remember that they should not be fed meat. While they will cheerfully eat meat the way they will just about everything else, the meat can rot in the tank, creating a bad smell or creating an issue where there is bacteria introduced into the environment.

If you notice that your cockroaches are not feeding well, remember that you should check the temperatures. Typically the higher the temperature, the more ravenous the cockroaches will be. Cockroaches that are kept in environments that are too cool will be sluggish. Cockroaches are considered to be cold-blooded animals, and they need the heat to spur their digestion.

Also remember that cockroaches will eat more the older they get. If you get a fair number of cockroach nymphs, you may notice that they eat a lot right after you get them, but then their food consumption will slow down. This is entirely normal. Nymphs were often not fed very much when they were first shipped, which prevents a build up of droppings and smell in the shipping container. Then, when they first come out, they are often much warmer than they were before, which kicks their appetite into high gear. This is something that can make a huge difference to the amount of appetite that they have.

Do not worry if you have cockroaches that do not eat a lot right away. Wait until they have a few molts, and then watch them eat. Most cockroaches will also be very uninterested in food until they molt, and after they molt, they will eat a lot.

Some companies have created their own mixes of cockroach feed, which is typically a dry mixture with plenty of different ingredients in it. There is nothing wrong with picking up some "roach chow," but you should always make sure that you are getting it from a reputable dealer. Find out what is in it before you offer it to your own cockroaches. Even if you purchase roach chow for your pets or your breeders, occasionally supplement their diet with slices of apple and orange to keep them hydrated.

11) Gut Loading

Gut loading is the process where people who are going to feed their cockroaches to other animals make sure that the feeder animals are full of nutrients to pass along. Simply feeding your animals cockroaches that have been fed on a subsistence diet will not actually give your animal that many benefits. In fact, it is often like eating potato chips, where all you really get are empty calories.

When you are looking for a good way to get your pets the food that they need, the best option is always for you to feed your own cockroaches very well. This usually means a diet that is between 40 to 50 percent protein of one sort or another and the rest made up in dark green vegetables and colorful fruits.

Use dry kibble of one sort or another to make up the protein, and use your own vegetable scraps to make up the rest. This is something that can offer your pets a lot more nutrition than simply feeding them crickets or cockroaches that come straight from the pet store or right out of the shipping box.

One thing that many people do when they gut load their insects is to think about calcium. Many reptiles need more calcium than they get, but unless you are very careful and diligent, it can be

76

hard to get enough calcium into your feeder insects.

When you want to make sure that you are feeding your insects something that is high in calcium, consider picking up a calcium supplement. A calcium supplement from the pet store is generally offered in a powdered form. After you buy it, you can simply sprinkle it over the food that you already give your cockroaches.

If you are not planning to feed your cockroaches to another animal, you should not bother with the calcium. A small amount is harmless, but if you give them too much, you will quickly discover that they can develop issues with their legs or even acquire a condition where their shells are too thick and immobile.

When you are looking at gut loading cockroaches for your lizards or snakes, you will discover that it is best if you simply feed them well and then add calcium to their diet once every two weeks. This is usually enough to get the benefits to your reptiles.

When you go to feed your lizard or snake some cockroaches, most of the time you can simply drop the cockroaches into the enclosure with the animal. Then instinct will take over and your animal will hunt the cockroach as it would in the wild.

However, you will find that some lizards and snakes are a little slower than might be preferred when it comes to hunting cockroaches. If this occurs, pop your cockroach into a resealable plastic bag and leave it in the freezer for a few minutes. When you do this, you are slowing the cockroach down so that it is a little harder for the cockroach to get away.

If your lizard or snake has not eaten the cockroach in 20 minutes or so, remove the cockroach and try again later

Chapter 6) Breeding

Whether you are interested in creating a thriving feeder colony of cockroaches for your pets or you simply would like to have more cockroaches around, you will find that breeding cockroaches is a fairly simple thing. Cockroaches are highly prolific animals, and just about all of the common species that you are likely to be raising are easy to sex and breed.

1) Should I Breed?

While breeding cockroaches is quite simple, you will find that you may start wondering whether you should breed them or not. Breeding leads to increased animals to care for, and before you get into that space, you should figure out why you are doing it.

Do not breed simply because you want more animals around. Be aware of the number of animals that you need and do not exceed that. Also, look at your own ability to care for the animals in question and go from there. If you take on more animals than you can care for, things can go quite badly for you.

Also remember that your household might not be able to handle as large a colony as you might want. If you have someone in the house with an insect phobia, it might be best to keep your colony down.

Remember that although there are people who sell cockroaches for money, it takes a lot of time and a lot of effort to get there.

Always make sure that you know what you are getting into before you breed your animals. Cockroaches can breed very quickly once they get started so be aware of the costs and the time investment required.

2) Sexing Cockroaches

There are several different species that you may choose to breed, and there are different methods for sexing them.

For example, when you are dealing with Madagascar hissing cockroaches, the females are generally longer and fatter than the males. From time to time, you do see a smaller female or a truly gigantic male, but this typically holds true. Males also have longer horn-like structures that protrude from their head. These horns are used for dominance displays and combat with other males. Females also have horns, but they are far less pronounced. In general, the males are a very dark brown, one that is nearly completely black. While the females have black heads, their bodies generally shade more towards a truer brown.

Dubia cockroaches are the other type of cockroaches that you are likely to be breeding. When you are looking at the adults, you will find that, like the Madagascar hissing cockroaches, the females are both longer and fatter. On top of that, the males have long wings that stretch all the way back across their backs, nearly to the end of their abdomens. On the other hand, females have much shorter, much stubbier wings that leave their striped abdomens bare.

You can also sex dubia cockroaches before they are fully mature. Just take a look at the final segment on their abdomen, the one that is farthest away from their head. On males, this segment will be rather narrow and tapered. On females, this segment is nearly as wide as the full girth of the cockroach.

Understanding how to sex cockroaches can help you to figure out why you are breeding a lot of animals or nearly none at all. When you want a great breeding colony, it is always better to have the females outnumber the males.

3) Breeding

There are two variables that come into play with regards to breeding cockroaches. Once you have at least one healthy male and a healthy female, you need to adjust the temperature and the humidity.

Raise the temperature by around 10 degrees Fahrenheit. Do not go above 95 degrees or so, but in general, most cockroaches will start to mate at around 90 degrees, with tropical cockroaches needing a little more heat.

As you raise the temperature, check the humidity. If you want to be scientific about it, you can install a temperature gauge inside your tank. Aim for between 50 and 70 percent humidity. Once again, tropical cockroaches will require more humidity than those found in northern climes.

After you get the tank set up with the right environmental conditions, you will discover that breeding will just happen. There is no reason to remove the mother, the father or the other cockroaches in the tank to a different location. Adult cockroaches do not prey on smaller cockroaches of their own species.

After a certain amount of time, you will see that the female cockroach has produced an egg sac, known as the ootheca. Depending on the species, she might hide the egg sac or carry it around with her. In matter of days or weeks, you will start to see very small cockroaches in the terrarium.

Remember that they will be white at first, before they darken to their adult protective coloration. The white color is only because their shells have not hardened yet. You do not have albino cockroaches or cockroaches that are somehow defective.

This is all the knowledge it takes to breed cockroaches. As a matter of fact, most people realize that it takes a lot more effort to prevent cockroaches from breeding than to convince them to do it.

Once the young are born, consider whether you need to bring more food to the tank or whether you need to remove some cockroaches to a new tank to give the colony more space.

Chapter 7) Health Issues

Cockroaches are notoriously hardy, but it is definitely true that from time to time, there are health problems to take care of. While there is a certain amount of die-off that is to be expected in any kind of colony setting, you will find that observing your cockroaches and taking good care of them can help you prevent serious issues in the future.

1) Mites

Though most people consider cockroaches to be pests, it is also important to remember that these pests can have pests of their own! Hissing cockroaches especially are prone to mites, some of which might already be on the insect when you get it, and others that may arrive on the bedding that you use.

Mites are very small animals that hang on to the cockroaches. Though it was previously thought that mites were parasites, the truth is that they simply live on the cockroaches. They take small amounts of food from the cockroaches' meals, but they do not do anything to harm the cockroaches themselves. There are many different types of mites that may occur on cockroaches, but all of them are harmless to humans and other animals. All of them can be treated using the same measures.

Mites are recognized visually. They are usually just about the size of the head of a pin, and they can breed very quickly. If you find that there are mites on your cockroaches, it is best to act as soon as you see them, as otherwise they can multiply rapidly.

Getting rid of mites is a fairly straightforward process, but it is time-consuming. Leaving just a few mites in the tank can start the problem all over again.

For a significantly higher success rate, you should have another tank on hand that is sterilized and lined with substrate that you know is mite-free. If you do not have another tank, temporarily move your cockroaches to a jar with holes or some other holding facility as you clean their main tank, remove the bedding, and replace it with clean bedding.

The first method to getting rid of the mites is often called the "shake and bake" method. It is quite easy, and it can be used for a small to medium sized number of cockroaches. Get a large plastic bag and fill it with several teaspoons of white flour. Select a cockroach from your affected colony and drop it into the bag. Seal the bag, and shake the cockroach vigorously for at least twenty seconds. Then remove the cockroach. The flour sticks to the mites and pulls them off of the cockroach. Inspect your cockroach carefully for any signs of mites, and once it is clean, use a new plant mister full of clean water to get the flour off of your pet cockroach. Then place the cockroach into the clean enclosure. Repeat until all of your cockroaches have gone through the process.

The second method includes getting a can of compressed air. These canisters are typically designed to get dust and debris off of keyboards and mechanical items, and they can be found at any computer store. Set up a clean cage where the cockroaches can be placed, and then pick up a cockroach. Stand with the cockroach over the sink, and use the compressed air to blow the mites off of the cockroach itself. Then place the cockroach in the clean cage and repeat the process.

The third method that is frequently used to remove mites from cockroaches is a paintbrush. Pick up your cockroach and use a damp paintbrush to remove the mites. Then you can place the clean cockroach into the clean enclosure and rinse off the paintbrush before you repeat the procedure.

Resist the urge to pull the mites off of your cockroaches by hand. They are very small, and they are typically sitting on very delicate parts of your pets. Trying to remove the cockroaches by hand can damage your cockroaches' bodies.

Do not use rubbing alcohol to remove the mites. Some people recommend damping a cotton ball with rubbing alcohol and then using it to remove the mites, but this can actually be harmful to the roach itself.

Do not use any mite-killer to kill off the mites. Mite killers are chemical sprays or powders that are used to kill mites in homes and on other pets, but they are fairly likely to kill your roaches as well.

Some people also suggest using diatomaceous earth to deal with mites, but this is something that can also harm your cockroaches. Diatomaceous earth works by being small and sharp enough that it rips holes through the insects' shells. As you may guess, this will do the same thing to your cockroach colony that it will do to the mites!

Some cockroach keepers simply allow the mites to be. They state that the mites do not do any harm, and that they do not affect the cockroaches' lives at all. This is a decision that you have to make for yourself. Either of the methods mentioned above will not harm your cockroach, and many people believe that the removal of the mites improves the aesthetics of the animal. However, if you have a very large colony of cockroaches and there is no way you can remove all the mites, there is no harm in simply allowing the mites and the cockroaches to coexist.

2) Shedding Problems

Depending on the species and the general health of the animal, cockroaches can shed as many as five or six times during their lives. Shedding is the process where they grow larger, and they stop shedding once they reach adulthood. Shedding is such an important part of the lives of cockroaches that many experts list cockroach lifespans as being the number of years that they live after their shed. For example, the rhinoceros cockroach is said to have three or four sheds, and then after that, it lives for about five years.

Some cockroaches have problems getting rid of their old shells. If they cannot easily rip through the old shell, they may suffer issues relating to infection and deformity.

If you notice that a cockroach has become significantly duller or paler, that means that it is on the verge of shedding. If you do not see it shed within a few days, there may be an issue that you need to address.

One easy way to take care of the shedding issue is to simply raise the humidity in the enclosure. Spray the walls down with warm water in the morning. In most normal cases, by early evening, you will discover that the cockroach has shed completely.

Another easy way to get the cockroach to shed is to simply wet a paper towel and squeeze it until it is damp but not dripping. Roll the cockroach up in the towel and hold one end closed. The cockroach will nose around until it crawls out the other end. The moisture from the paper towel is often enough cause the insect to shed its skin.

If you notice that molting is an issue throughout your colony, it is time to look at the humidity requirements. Madagascar hissing cockroaches are tropical animals, and in many cases, they are used to dealing with between 75 to 80 percent humidity in their environments. Other cockroaches, ones from North America or the northern parts of Europe, often end up needing significantly less humidity.

When you think about it, simply spray the walls of the enclosure down with warm water, or drip some water into the substrate. A certain amount of dampness will not harm your roaches, and as a matter of fact, it will make many species thrive.

Under no circumstances should you try to remove a cockroach's partially shed skin on your own. A skin that has not been shed is still very tightly attached to the cockroach's body, particularly to its legs. If you pull too hard, you might pull off some legs or you might even kill the insect.

3) Escapees

No matter how careful you are, you will probably find yourself dealing with an escapee at some point or another in your cockroach husbanding experience. This is something that happens to the best keepers in the world, so do not stress out about it.

If you have an escaped hissing cockroach, you can often listen for it. The noise that a hissing cockroach makes is fairly distinctive, so on a regular basis, unplug noisy electrical appliances, like your refrigerator, and listen for it.

Fortunately, most of the cockroaches that are kept as pets and as feeders are relatively slow. If one falls out of your hand, you can frequently pick it up again very easily. If your cockroach falls out of your hand or jumps as you are handling it, stay calm, and do not jump up and down. Instead, simply step back so that you do not step on the cockroach and pick it up again. In general, the larger a cockroach is, the slower it is.

Fortunately, if you live in a colder climate, you will not be dealing with an infestation as a result of escapees. Cockroaches need a certain temperature to breed or even for the eggs to hatch, so that means that in general, you will be dealing with just one escapee, rather than one escapee and dozens of young.

This is one area where cockroaches are often cited as being preferable to crickets, as crickets will breed very quickly if they escape.

If you are searching for an escaped cockroach, do your best to look in a variety of low, tight corners, Check behind large appliances and furniture, and if possible, have someone standing behind you to see if the animal scuttles away from the area as items are moved.

This is something that takes some time and effort, but be patient and keep looking. People have found missing cockroaches months after they lost them. Cockroaches can go for quite some time without food, and you will discover that they are quite hardy- even if they are wandering around in a hostile environment.

If you have cats and dogs, however, it may be very hard to prevent them from eating your cockroach escapees.

4) Aggression

For the most part, your cockroaches should not be aggressive. In the wild, there may be some dominance displays and some biting due to mating issues, but for the most part, cockroaches do very well simply living together.

However, once in a while, you may find that your cockroaches suddenly turn on each other. This is something that is more common in large colonies, but it can also happen in small groups as well. It is possible that you will go to bed one night with a healthy tank and wake up the next morning to find that your cockroaches have torn legs, antennae, and wings off of each other.

When you are looking at aggression in what was previously a peaceful tank, chances are good that the issue is due to either food or temperature. Any change in the insects' environment, no matter how small, is something that can disrupt their behavior.

For example, while some people do choose to feed their cockroaches once a week, others state that this is simply not enough for a larger colony. Consider feeding your roaches more often and varying the amount of food that they get. Try placing the food in several different places in the tank, as some cockroaches might be fighting for territory in a certain area.

Another consideration is the temperature. Particularly high temperatures might mean that your cockroaches are having issues settling down. Check the temperature in the tank, and compare it to what your cockroaches would experience in their native habitat. If the temperatures are too high, you may want to dial it back a little.

Males tend to be far more aggressive than females, and there is some anecdotal evidence that states that colonies that have more females than males tend to be calmer. If you are dealing with a colony that seems to be relentlessly aggressive, try splitting it up, ensuring that the males are either living on their own or housed in groups that are predominantly female.

5) Feeding Issues

Sometimes, it looks like your cockroaches are not eating. This can be a big concern, especially if you are a new owner. It takes a fairly long time before your cockroach starves to death, but it can happen, so take a moment to consider what you can do to prevent this fate.

First of all, check the humidity in your tank. If it seems a bit dry, mist down the walls of the tank. This can encourage stubborn eaters to remember that they need food.

You may also try to tempt your stubborn cockroaches with food that they have not had before. When enticing anorexic roaches to eat, consider brightly colored and very sweet items, like apples, oranges and slices of bananas.

You should also look into your tank to see if there is something in it that the cockroach is eating instead. Cockroaches are extremely durable, and in some cases, they may have decided that they would rather eat something like cardboard or paper. While there is nothing wrong with them eating these substances, it might be keeping them from the food that is better for them. This is quite a concern if you are raising cockroaches to feed to your other pets!

If you keep your cockroaches on paper products or if you give them toilet paper rolls to play with, they may be chomping on those items instead. Remove them from the tank and see if the cockroaches start eating.

Finally, remember that cockroaches are small. The amount of food that they eat is negligible to humans, and if you only have two or three cockroaches, there is a chance that you will never notice the amount of food that they eat.

Also don't forget that some cockroaches simply seem shy. They do not want to eat if there is a light on, and they might run and hide when you are around. There is nothing wrong with this behavior. It is what will keep your pets alive and well if they lived in the wild.

When it comes to taking care of your cockroaches and whether or not they are feeding, remember that there are some nymphs that simply never eat. Evolutionarily speaking, cockroaches thrive on a system where they simply have many, many babies instead of having just a few that are very hardy. This means that there are more animals that simply do not go on to mature and thrive.

As long as the cockroaches themselves seem to be in good shape, there is nothing wrong with simply letting them be. If you are just raising a few cockroaches and they do not seem to be feeding, try switching around their food. Most cockroaches are fairly indiscriminate when it comes to what they eat, but some keepers swear that their pets play favorites!

6) Limbs Dropping Off

It is fairly distressing to see limbs or antennae dropping off of your cockroach. This is something that does happen from time to time, and unlike some other animals, these parts will not grow back. However, it is worth noting that cockroaches are very tough, and that in many cases, they will get along just fine without the given parts, though there is a chance that they will no longer be able to climb as well as they once could.

If you notice that your cockroach has limbs dropping off, the first thing to consider is the humidity. The cockroach might be getting enough to drink, but if there is not enough ambient water in the air, the animal might be having issues with its shell.

Adjust the humidity by spraying water on the walls or by making sure that the substrate that the cockroach lives on is moistened. You might even choose to wet a paper towel and leave it in the enclosure.

Keep an eye on the cockroach. If it is having a hard time living in a colony, you might decide that it should live on its own for the rest of its life. Isolating a cockroach that is weaker than the others can prolong its life and allow it to live without harassment. At the very least, if the cockroach is living in a small isolated cage, it can easily get to the food and water without needing to fight for access.

If you notice that a cockroach is dropping limbs, keep a close eye on it and make sure that you know how well it is eating and drinking. A change in the humidity should fix most things, but you should always be willing to try something else if you need to.

Chapter 8) Cockroaches in Your Home

Whether devoted cockroach keepers like it or not, cockroaches are considered pests by most of the world. Given the issues that they can create when they swarm a human habitation, it makes sense that some people would be a little worried about having them around.

While keeping in mind that the cockroaches found in houses are far different from the ones that you keep as pets or feeders, it is worth knowing a little bit about how to prevent cockroaches from invading your home.

1) Preventing Cockroaches in Your Home

There is a huge difference between insects that you keep in a tank and tend carefully and those that come into your home uninvited. Cockroaches that invade homes should be dealt with very quickly for a number of reasons, and even if you notice just one or two, you need to take action.

Cockroaches that come into your home from the outdoors or from other houses are disease carriers. They crawl through extremely unsanitary conditions, and they also crawl through toxins. Cockroaches are notoriously tough, and just because they can carry toxins on their back does not mean that those toxins are safe for you, your family or your pets.

When you are dealing with wild cockroaches, remember that the best way to prevent them is to keep them away from any kind of food. Cockroaches go where there is food, and if they discover that there is none, they will move on.

However, given the fact that cockroaches can survive, if not thrive, on very small amounts of food that are found on a very random basis, this can be easier said than done!

If you are dealing with a cockroach issue, start by cleaning your home as thoroughly as you can. Sweep and vacuum on a regular basis. You may choose to sweep the floors once a day or once every few days, and you should vacuum at least once a week. This takes care of the crumbs that might be an issue.

Go through your pantry and remove everything from the cardboard boxes. Cockroaches can find their way into boxes that have been sealed after being open, and there is something very disturbing about finding live or dead roaches in your cereal. Instead, invest in plastic canisters with air-tight lids. These canisters can be snapped shut to keep the cockroaches out of them.

You may also choose to use glass jars, but if you do this, make sure that there is a rubber seal around the edge of the jar lid, or that the jar has a screw lid. A lid that simply rests in place is just less secure.

People often notice that cockroaches end up in their drawers and their cabinets, and what you need to understand is that they are not coming in through the doors. Cracks in your cabinets and drawers are hard to notice due to the fact that there are usually a lot of things in the way. However, a jumble of kitchen utensils and food containers are not going to get in the way of a determined cockroach.

To keep insects out of your drawers, look into using contact paper to seal the cracks. Contact paper is a type of paper that is sticky on one side and slick on the other. The sticky side is applied to the drawer or the cabinet, lining it and covering up whatever cracks may be there. This is an effective solution that manages to give your cabinet or drawer a little bit of visual interest as well.

Because you may go a day or two without emptying your trash, make sure that you purchase trashcans with sealed lids. Lids can make sure that cockroaches cannot enter your trash and eat the scraps that they find there.

Make sure that you do the dishes as soon as they are dirty. Do not leave them in the sink overnight, as that is when cockroaches are the most active. Ideally, you will do the dishes and then wipe down the sink when you are done. This is something that can make a huge difference in many different kinds of cockroach problems.

Another thing for you to consider when you are looking at keeping cockroaches out is wiping down all of your wet surfaces. Cockroaches are typically animals that are drawn to water and to humidity, and this means that they are going to be drawn to your kitchen and your bathroom. After you use the shower or the sink, wipe it dry with a paper towel. This helps reduce the amount of moisture around.

A good way to prevent cockroaches from coming in from the outside is to cut back the hedges and any undergrowth that might be growing up to your home. Cockroaches, like a wide variety of other pests, need cover before they can travel from place to place. This is something that actually makes them quite shy. If you cut back the brush around your home, you will make it much less likely for them to cross into your home.

Do not handle any wild cockroaches in your home if at all possible. The less contact you come into with them, the better. Some people choose to kill the isolated cockroach with their shoes, but even this is not a great idea, as this can transfer eggs to the shoe, which can then hatch! Instead, it is often a good idea to simply capture the cockroach in a paper cup or some other container and then to flush it down the toilet. This prevents a mess, and it also gets rid of any eggs a female might be carrying.

Cockroaches can flatten themselves out a great deal. This means that if they are trying hard enough, they can come into your home through a number of different ways. If you want to make sure that your home is as secure as possible, you need to seal up as many gaps and cracks as possible.

Grab a container of silicone caulk, and go through your house as thoroughly as you can. Wherever you see a gap, fill it with caulk. Be thorough, and make sure that every gap gets filled. This is something that you can easily do in an apartment, where you may not be able to control what other people do in their own space and whether they can keep it clean.

2) Killing Cockroaches

If you need to kill roaches in your home, one of the first methods that you should consider is diatomaceous earth. Diatomaceous earth is made up of very small, very sharp granules. When they are sprinkled in the places where cockroaches travel, they rip open the roaches' bodies. The benefit to this material, however, is that it is very safe to use in just about any house. There are types of diatomaceous earth that are even considered food safe!

Simply find a container of diatomaceoous earth and sprinkle them in places where you have seen roaches. This may make your home look dusty for a while, but it can really make a difference to the situation.

If your issue is getting out of hand, it is time to talk to an exterminator. An exterminator, also called a pest control specialist, is an individual who will have a number of different methods for dealing with an infestation of roaches. You should always talk closely with your exterminator and ask them what they intend to do.

When you are talking about your issue with your exterminator, it is ideal if you have a sample of the insect in question to offer. As you know, there are many different types of cockroaches out there, and the exterminator will be in much better shape if they know what kind they are dealing with.

There are different methods for dealing with American cockroaches versus German cockroaches versus oriental cockroaches, so do your best to catch a specimen and keep it in a jar to show the exterminator.

Always make sure that your exterminator is aware of the condition of the people in the house. For example, he or she should be made aware of any children in the house, any elderly or any people with disabilities or health problems. In addition to that, they should also be made aware of any animals that you might have, especially if these animals are delicate, as in the case of fish, reptiles or even pet roaches!

A good exterminator will offer you a projected schedule of treatment and also a cost for that treatment. Most exterminators will offer at least some kind of guarantee. Remember that there can be an enormous range of charges for different treatments, so get a second opinion.

You are allowing the exterminator into your home to spray potentially harmful chemicals into the area. You should make sure that they are someone that you trust. Interview the exterminator as closely as you need to, and do not rest until you have gotten the information that you need.

If you are dealing with a serious roach infestation, it is important to hand the issue over to a professional. The more serious the issue, the more potent the measures needed to fix it, and these measures should not be handled by amateurs.

For example, never set off a "bug bomb" in your house. A bug bomb is essentially a cloud of toxins that are designed to get rid of insects, but there are a few flaws. In many cases, you are simply introducing harmful chemicals into your home in an uncontrolled fashion.

On top of that, these bug bombs will often drive the insects away, but they will not drive them very far. In many cases, you will simply get cockroaches that have been pushed to other rooms of the house. This can be a serious problem because it can actually drive the insects to areas that were previously cockroach free.

Also avoid using anything that is labeled 'roach chalk'. Essentially, these products state that you can draw a line around the edge of your rooms and that roaches won't cross them. In the first place, this is not true. Cockroaches will absolutely cross these lines, and then they will spread the toxins in the chalk around the house. This product is often sold in grocery stores and at truck stops. Just skip them entirely.

When you are looking to get rid of cockroaches, remember that anything that sounds too good to be true, generally is. Too many people have miracle cures for cockroaches out there, but the truth is that cockroach infestations are complex issues that usually require a fairly complex solution.

Talk with plenty of professionals, be willing to pay what the service is worth, and educate yourself on the problem in your home.

3) A Word About Apartments

Cockroaches are a serious problem in many multiple-family dwellings, and in many cases, the more transient the population, the more likely it is that cockroaches have been brought in from one place or another.

If you live in an apartment, you will discover that cockroaches can, in some cases, become a fact of life. It does not matter how clean you are or how tidy you are, you are going to discover that having a neighbor with less than great habits can drive cockroaches into your living space.

There are a few things that you can do to protect yourself when you are looking at making sure that you can deal with cockroaches.

First, always ask your management what their pest control regimen is. Have they dealt with pests in the past? If they have, do not hold it against them. Instead, judge them by what they are willing to do in the future.

Is there a clause that allows you to move out with a penalty if there is a serious infestation issue? Are they willing to offer you reduced rent or to move you to another apartment if things get bad?

If you are moving out of a place with cockroaches, it is worth your while to be careful about what you bring with you to your next apartment. You can absolutely carry cockroaches from one living space to another, so pack with care.

Make sure that you inspect anything that you are bringing along, and then wrap it securely. Do not leave any gaps in your packaging that will attract cockroaches, and always inspect your boxes before you proceed to pack things into it.

Chapter 9) Eating Cockroaches

If you've read this far, you know that plenty of animals eat cockroaches, but did you know that people can eat them, too? While people will eat just about anything if they get hungry enough, the truth is that there are plenty of people out there who will eat cockroaches as part of a meal.

Historically, cockroaches have been a source of food for people who lived largely off of the land. They are widely considered too much trouble when it comes to cultivation in any realistic sense, but they were always good for a snack when times were lean.

Cockroaches do not provide much in the way of nutrition for a person. While they are a fantastic source of food for reptiles of all kinds, there is just not enough of nutrients in them to satisfy a human unless they are eating in very large quantities. With that in mind, they can add much needed protein to a person's diet, just as they do for reptiles.

The first thing to consider is that cockroaches should not be eaten out of hand. Cockroaches that you find crawling on the floor are not fit to be eaten, as they are usually dirty and in some cases, covered with pesticides and toxins.

The cockroaches that can be eaten are bred and grown for that purpose. Because most insects taste slightly of the food that they have eaten themselves, they are often fed on high-quality feed like grain, fruits and vegetables.

Another option that is considered is the size. While there are very large cockroaches out there, most are small and not really worth eating. When it comes to both experimental cuisine and traditional cuisine, larger is better, allowing for both more nutrition and more crunch. As a result, it is very clear to see why

Madagascar hissing cockroaches have ended up on the menu so often!

Cockroaches that are designed to be eaten by humans are raised just like any other food animal. Ideally, they are kept in sterile conditions, they are not dusted with anything harmful to people, and they are fed well.

In some cases, they are skewered and cooked over a high heat, often with oil drizzled on top to improve the flavor. This is one fairly traditional method for cooking cockroaches, and it is quite similar to a Japanese famine-time custom of eating skewered cicadas.

On the experimental food front, cockroaches are typically fried, though they may be soaked in flavoring agents or otherwise infused with flavors or liquids before they are cooked. This is something that can greatly enhance the flavor of the insects, which is otherwise fairly dry and uninteresting.

In some cases, cockroaches are fed interesting foods before they are consumed. One Norwegian food lab fed their cockroaches on chamomile, black currants and split peas before they were sliced and sautéed.

It is generally recommended that you do not eat cockroaches yourself, even if you are the one that raised them. In 2012, a Florida man died of asphyxiation after winning a cockroach eating contest at a local pet store.

Experts stated that the legs from hissing cockroaches, which were the kind being eaten, as well as the kind that is most commonly used in foodie experiments, are sharp enough to cut, suggesting that eating the legs would be a little like eating fish hooks.

Cockroaches that are designated for human consumption are very carefully prepared. Any parts that are very sharp or dangerous are carefully snipped off, and the cockroaches themselves are inspected for any sign of damage or contamination.

If you are curious about eating cockroaches, it is best to wait for the nearest foodie event to serve them. Several types of establishments might serve cockroaches. You may find them at a daring dining establishment, or you may discover them on the menu when you are looking at a banquet for sustainable eating. Cockroaches are often used to urge people to look beyond the basics when it comes to food. As this very guide can tell you, they are easy to raise and easy to breed, so why should we limit ourselves to beef, chicken and ham?

The next time that you start thinking about cockroaches, think about the fact that they are animals. Try divorcing them from the associations that we have been given regarding them, and consider what other uses they could be put to.

Resources

http://www.roachforum.com/
An excellent place for cockroach breeders and hobbyist.

http://www.blattodea.net/
An excellent resource for first-time cockroach owners.

http://www.theroachhut.co.uk/
UK source for several different species of cockroaches.

http://www.roachcrossing.com/
US source for feeder and pet cockroaches of all sorts. They also do their own research and are an invaluable resource for first-time cockroach owners.

http://www.petco.com/product/5914/Petco-Pet-Keeper-for-Small-Animals.aspx
This Kritter Keeper in large is a great enclosure when you are looking at keeping just a few cockroaches.

http://www.petsmart.com
Large US chain which always has bigger aquariums.

http://www.swelluk.com/
Excellent UK aquarium source.

Sources for pictures used in this book:
http://commons.wikimedia.org/wiki/File:Snodgrass_common_household_roaches.png
http://commons.wikimedia.org/wiki/File:Madagascar_hissing_cockroaches_bugs_gromphadorhina_portentosa.jpg